Praise for Moderating Usability Tests

Interacting with participants in a calm and neutral manner may well be the most difficult part of doing usability testing. Now you no longer have to worry about how to do that. Just follow Dumas and Loring's wonderful, practical advice and you will be prepared not only for typical encounters, but also for the unusual and unexpected, for doing remote testing, and for working with special populations. *Moderating Usability Tests* is a great resource for anyone who interacts with usability test participants.

—Janice (Ginny) Redish, President, Redish & Associates, Inc.

Everyone talks about research methods, but the formal aspects of those methods only get you so far. The difference between getting a little data or a lot of data, only discovering problems or getting ideas about solutions, bias or validity, throw-away data versus generalizable insights, often depends on the soft skills—the ability to effectively moderate testing. In the past, you were expected to get these skills through apprenticeships or trial and error. *Moderating Usability Tests* removes the mystery and provides practical advice on how to get the most out of research. It will be invaluable to students learning about usability testing for the first time, people newly charged with evaluating products, and even old hands looking to refine and improve their technique.

—Arnold (Arnie) Lund, Director of User Experience, Microsoft

You may not think that being a "Gracious Host" is among your assignments in moderating a usability test, but you will learn why this and other roles with similarly illuminating names are important to your success. In this generous book, Dumas and Loring give the benefit of their decades of experience and astute observation of both the foundational and the subtle aspects of conducting usability tests. Many questions you didn't think to ask until you were on the hot seat are answered here, and will help you achieve a level of confidence as a test moderator that may have seemed beyond reach, even if your participants are from challenging-to-test populations. With this highly ethical and thoroughly grounded program for developing moderator skills and avoiding pitfalls, Dumas and Loring make a strong contribution to the body of knowledge on testing products. The big surprise of the book is that their clear, reasoned, and detailed suggestions about interacting with test participants and developers will likely spill over and improve your relationships with coworkers, family, neighbors, and friends.

—Elisabeth Bayle, Bayle Collaborations

At this point, virtually everyone in the software industry knows what usability testing is. An unfortunate side effect of this awareness is that many people are conducting usability testing who have no idea how to do so in a way that will yield valid, reliable, and useful data. Other than the design of the test itself, proper and effective moderation of test sessions is one of the most important—and least understood—aspects of usability testing. Here is a book by two highly regarded experts that covers this topic thoroughly in a very readable format. No one who has not already been well trained should attempt to conduct usability testing without first reading this book cover to cover, and viewing all the excellent videos the authors provide on the book's web site.

—Deborah J. Mayhew, Deborah J. Mayhew & Associates

Moderating Usability Tests

The Morgan Kaufmann Series in Interactive Technologies

Series Editors: Stuart Card, PARC; Jonathan Grudin, Microsoft; and Jakob Nielsen, Nielsen Norman Group

Moderating Usability Tests
Principles and Practices for Interacting

Joseph S. Dumas

Beth A. Loring

ELSEVIER

AMSTERDAM • BOSTON • HEIDELBERG • LONDON
NEW YORK • OXFORD • PARIS • SAN DIEGO
SAN FRANCISCO • SINGAPORE • SYDNEY • TOKYO
Morgan Kaufmann is an imprint of Elsevier

MORGAN KAUFMANN PUBLISHERS

Publisher	Denise E. M. Penrose
Publishing Services Manager	George Morrison
Project Manager	Marilyn E. Rash
Assistant Editor	Mary E. James
Copyeditor	Joan Flaherty
Proofreader	Dianne Wood
Indexer	Ted Laux
Cover Design	Pam Poll
Typesetting/Illustration Formatting	SNP Best-set Typesetter Ltd., Hong Kong
Interior Printer	Sheridan Books
Cover Printer	Phoenix Color Corp.

Morgan Kaufmann Publishers is an imprint of Elsevier.
30 Corporate Drive, Burlington, MA 01803

This book is printed on acid-free paper.

Library of Congress Cataloging-in-Publication Data
Dumas, Joseph S., 1943–
    Moderating usability tests : principles and practices for interacting / Joseph S. Dumas, Beth A. Loring.
        p.   cm.—(The Morgan Kaufmann interactive technologies series)
    Includes bibliographical references and index.
    ISBN-13: 978-0-12-373933-9 (alk. paper)   1. Human–computer interaction.   2. User interfaces (Computer systems)—Testing.   I. Loring, Beth A.   II. Title.
    QA76.9.H85D858 2008
    004.01'9—dc22   2007046074

For information on all Morgan Kaufmann publications, visit our
Web site at *www.mkp.com* or *www.books.elsevier.com*.

Printed in the United States
08   09   10   11   12   10   9   8   7   6   5   4   3   2   1

Contents

Preface

From many conversations with other usability testers, we know that our training as moderators was typical. We both learned how to moderate usability tests the same way: a colleague let us watch a few sessions and then watched us struggle for one or two sessions. There was no formal training and no set of professionally accepted procedures.

Since then, we have moderated numerous usability tests and watched thousands of sessions and debated good and bad practices with colleagues over the years. Clearly, a logical and practical approach to training moderators to be as effective as possible in the pursuit of valid usability testing was overdue. We looked at the literature and found some good, but limited, advice. Two currently available books on how to conduct a test, Dumas and Redish (1993) and Rubin (1994), each have one chapter on moderating. A later source, Snyder (2003), has several chapters, but they focus on moderating tests of paper prototypes. No previously published authors have presented a set of rules that underlie effective moderating.

While preparing a tutorial on test moderation for the Usability Professionals' Association annual meeting, we realized that our best practices (developed through the trials and errors of two decades) could be captured in a set of professional guidelines. Moderating is more than just an art. Unfortunately, it is not something any intelligent person can walk in and do effectively. So, we created the ten golden rules of moderating that we present in chapters 3 and 4 and apply throughout.

Our experiences made very clear that new moderators need more than just verbal descriptions of how to interact with test participants. They need to see experienced moderators in action, dealing with specific situations. To illustrate how the rules translate into practice, we filmed a set of videos, several of which were later refilmed to accompany this book. In addition, we invited some of our colleagues to discuss what they saw in each video. We filmed these panel discussions, too. The videos and panel discussions can be found on the publisher's web site (*www.mkp.com/moderatingtests*). Also on the site is a table that lists all of the places in the book where each video and the web site are discussed.

Our goal for this book is to enrich the learning experience for new test moderators. In doing so, we hope to establish a consensus that we and our colleagues can use to move moderating from an art passed on in private to a set of agreed-upon practices that can be used in an effective training program.

Acknowledgments

We have had the support and hard work of many colleagues in creating this book. Several of our colleagues at Bentley College helped us clarify our ideas and our writing. Lena Dmitrieva and Eva Kaniasty reviewed the first drafts of the chapters, stimulating us to rethink some of our initial ideas. Rich Buttiglieri handled the video and recording technology and acted as the director for the videos. Andrew Wirtanen edited and transferred the videos onto the web site and Steve Salina filmed the panel discussions. Chris Hass's expertise in interacting with disabled participants provided much of the practical advice found in chapter 10.

The reviewers of the manuscript were an enormous help: Whitney Quesenbery, Kelly Gordon Vaughn, Catherine Courage, Ron Perkins, Steve Krug, and Mary Beth Rettger. Each one made a unique contribution that improved our ideas and writing. In addition, they were uniformly encouraging, which all authors need during the stressful days when a finished manuscript seems unreachable.

The colleagues who acted in the videos allowed us to illustrate our golden rules. We often asked them to portray poor practices, which they did with good humor. The expert panel we asked to comment on the videos provided exactly what we wanted—their spontaneous and honest reactions to what they saw. We had great fun watching them provide insights to supplement the videos and the book.

Finally, we want to thank Diane Cerra at Elsevier, who encouraged us to write this book and whose encouragement and guidance in its early stages were invaluable.

*   *   *

For seven years I taught a course in usability evaluation in the Bentley College graduate program. A section of every course included a discussion of moderating a test. I am grateful to the students in those classes who pushed me to examine my assumptions and to find ways to explain my thinking. My wife, Martie, has encouraged me to write this—my third book—without complaint. Somehow, back in high school, she saw the potential in me that I couldn't see in myself, and I saw in her the woman to spend a life with. Forty-three years later, our relationship continues to evolve.

Joe Dumas

The person to whom I am most grateful is Joe. He became my mentor way back in 1986 when usability testing was in its infancy, and our professional paths have crossed so often that they seem to be interwoven. It was Joe's idea to present a tutorial on the topic of interacting with test participants in the first place, and it was his idea to ask Diane Cerra about turning it into a book. He and I make a good team, and I am proud to call him my friend. I'm also grateful to my husband, Jon, and my son, Stephen, for giving me the time, space, and quiet to work on the book for so many months.

Beth Loring

About the Authors

Joe Dumas is a recognized expert in usability evaluation with twenty-five years of experience. He has moderated or observed others moderate thousands of usability testing sessions and has taught numerous students and usability professionals how to moderate. He is currently a usability consultant for Oracle Corporation. He was a senior human factors specialist at Bentley College's Design and Usability Center and taught graduate courses in the college's graduate program in Human Factors in Information Design. He has a Ph.D. in cognitive psychology from the State University of New York at Buffalo. He is the author of *A Practical Guide to Usability Testing* (with Ginny Redish), *Designing User Interfaces for Software*, and numerous articles written for researchers and practitioners.

Beth Loring is director of the Bentley Design and Usability Center, a research and consulting organization in Waltham, Massachusetts, USA. Beth has more than twenty years of experience in human factors engineering and product usability.

Since the mid 1980s, Beth has evaluated a wide array of products and services including desktop software, consumer products, web sites, business applications, and medical devices. Beth's online experience includes testing of financial, e-commerce, educational, intranet, and business-to-business applications.

Before joining Bentley, Beth was a principal research scientist at the American Institutes for Research in Concord, Massachusetts, and before that she served as the human factors team lead at IDEO Product Development in Boston. She teaches courses on usability testing in Bentley's Information Design Certificate Program.

Beth holds a master of science degree in engineering design from Tufts University and is a Certified Human Factors Professional. She has published more than twenty papers and is coauthor of a case study in *Understanding Your Users: A Practical Guide to User Requirements* (Morgan Kaufmann, 2005).

Moderating Usability Tests

Introduction

This book grew out of a tutorial that we presented at several Usability Professionals' Association (UPA) conferences starting in 2004. The success of the tutorial and the positive reaction of the participants led us to expand the scope and publish it as a book. It is intended for both newcomers to usability testing and experienced practitioners who want to reflect on their own practices.

1.1 **WHY THIS BOOK?**

Usability testing is now accepted as the evaluation method that influences product design the most (Rosenbaum, Rohm, & Homburg, 2000). To a large extent, successful usability testing depends on the skills of the person moderating the test. Most practitioners, however, learn how to moderate with little formal training and almost no feedback on their technique. They watch a few test sessions, moderate a few under supervision, and then proceed without further evaluation. It's not surprising that researchers have found that moderators, even in the same organization, don't follow the same practices (Boren & Ramey, 2000).

With the exception of short sections in Rubin (1994) and Snyder (2003), the literature on usability testing provides only general advice about interacting with test participants. In addition, there are no sources about either general principles or specific guidelines that provide a rationale for what moderators should or shouldn't do. Maybe this is because many people consider it part of the "art" of testing—a skill that is difficult to teach. The lack of relevant literature became a problem as we tried to teach people how to moderate. Therefore, in this book we go into depth about what to do (and what not to do) while interacting with test participants, and we present a set of rules of good practice.

Although one paper discusses situations that happen infrequently (that is, Loring & Patel, 2001), this book focuses on the common situations you will encounter in typical usability test sessions.

Over time, as we taught people how to moderate test sessions, we recognized the need to address the difficulties that moderators have with the following issues:

- learning how to overcome the initial anxiety that all new moderators experience
- staying in control of the test session
- being friendly but resisting becoming friends with participants
- dealing with personal—rather than the participant's—anxiety while struggling with usability problems
- balancing the trade-off between respecting participants' rights and pushing them to keep working
- knowing when and how to provide assistance to participants
- knowing how to probe for more information in an unbiased way

We hope that by addressing these and other issues in this book, we are making a valuable contribution to the field.

1.2 WHAT IS USABILITY TESTING?

Usability testing is a systematic way of observing actual and potential users of a product as they work with it under controlled conditions. It differs from other evaluation methods (such as quality assurance testing and product demonstrations) in that users try to complete tasks with a product on their own, with little help. Usability testing can be conducted in a laboratory, in a conference room, in the participant's environment, or remotely. Companies use this method to evaluate software, hardware, documentation, web sites, or any product with a user interface.

People are recruited, and usually compensated, to participate in the sessions. The goal of the study may be to uncover as many usability "bugs" as possible or to compare the usability of two products. A typical test session involves one participant and one moderator, lasts one to two hours, and includes these tasks:

- greeting the participant
- explaining the participant's rights and having him or her sign an informed consent
- explaining how the session will proceed
- guiding participants through a set of carefully selected tasks using the product, usually while thinking aloud
- recording data in one or more ways
- asking participants to summarize their experience

1.3 **THE IMPORTANCE OF MODERATING SKILLS**

Usability testing is like many things—easy to do, but hard to do well, and improved only through practice. The moderator's manner of interaction with participants from first recruiting them to thanking them at the end of the session is critical to the success of the test, the validity of the data, and the reputation of the moderator's organization. A test moderator must be unbiased and neutral regarding the product while being open and approachable to participants. This mixture of neutrality and approachability can be difficult to accomplish.

1.4 **THE GOLDEN RULES OF MODERATING**

Because most practitioners have learned only by example, they have seldom thought about the underlying principles of interacting with test participants. This led us to create the golden rules of interacting for moderators of usability tests. Our rules attempt to capture the best practices of interacting and provide a rationale for the practices described in the rules. The golden rules are:

1. Decide how to interact based on the purpose of the test.
2. Respect the participants' rights.
3. Remember your responsibility to future users.
4. Respect the participants as experts, but remain in charge.
5. Be professional, which includes being genuine.
6. Let the participants speak!
7. Remember that your intuition can hurt and help you.
8. Be unbiased.
9. Don't give away information inadvertently.
10. Watch yourself to keep sharp.

These rules will help you deal with new situations that you may encounter. We discuss these rules in detail in chapters 3 and 4.

1.5 **CULTURAL POINTS OF VIEW**

As we developed the material for this book, it became clear that we have the perspective and assumptions of usability specialists from the United States. In the book, we don't pretend to be otherwise. Instead, we state our biases up front and, where possible, discuss the need to be sensitive to cultural differences. In some cases, we cite literature about ways people from various cultures might react, but we don't attempt to guess how moderators and participants from other cultures might react. While we cite the few studies that describe ways that people from other cultures react to testing situations, we don't extrapolate beyond those studies by guessing how people from other cultures may react to our golden rules. The role of culture in usability testing is a new and interesting topic that is just beginning to be explored. It would be wonderful if practitioners from other cultures would create localized sets of golden rules!

1.6 **ABOUT THE SIDEBARS IN THIS BOOK**

Throughout the book, we offer sidebars to supplement the text. There are two types of sidebar. The first type presents short descriptions of relevant research studies that shed light on the topic at hand. Entitled "What the Research Says," each focuses on the practical implications of one published study.

The second type presents interviews with a variety of moderators at different organizations to allow them to express their views about a topic. Some of the moderators are new to moderating and others are quite experienced. Hence, these are entitled "Interviews with a New [or Experienced] Moderator." We found that stating the principles in moderators' own words can be more informative and colorful than if we described them ourselves.

1.7 **ABOUT THE VIDEOS THAT ACCOMPANY THIS BOOK**

Perhaps one reason that interacting with test participants has not been adequately addressed in the literature is that it doesn't lend itself well to description only on paper. To fully grasp both the art and the craft of test moderation, it helps to observe the interactions—including facial expressions, body language, tone of voice, and numerous other factors—that can have a direct impact on the test outcome but are difficult to convey in words.

To provide another dimension, we created a number of short videos to accompany this book. The role-playing videos show both good and not-so-good moderating practices. The filmed discussion by a panel of usability experts gives their opinions about what they see in the videos and discusses trade-offs that you might consider in each situation.

The videos, available on the book's web site were filmed in the Bentley College Design and Usability Center laboratory and were carefully scripted to illustrate the points we want to make.

Chapter 11 describes the type and content of the videos and suggests how to get the most value from them.

1.8 **ABOUT THE COMPANION WEB SITE**

The companion web site for this book is *www.mkp.com/ moderatingtests.* In addition to the videos, you will find downloadable versions of the sample test materials in the book, such as the checklists, forms, and recruiting scripts.

Chapter 2

Getting started as a test moderator

This chapter provides some guidance to moderators who are just starting out, but our comments may also be useful for experienced moderators who need to refresh their memories or who are training others. We discuss the attributes of a great moderator, types of testing, moderator (sometimes conflicting) roles, the basics of running a test, and finally some ways to get started quickly as a moderator. If you're familiar with usability testing, most of this material will be a review.

2.1 **WHAT MAKES A GREAT MODERATOR?**

Moderating usability tests is not as easy as it looks. If you're reading this book, you have probably seen a number of test sessions and may even have moderated some (or many) yourself. Some moderators make it look easy and others are so nervous that you feel nervous just watching them. Some moderators are able to elicit wonderful insights from test participants; others simply go through the motions. Some remain neutral and others seem to lead or bias the participants. Why is there so much variation?

Becoming a great test moderator takes four things:

1. Understanding the basics of usability testing
2. Interacting well with test participants (our golden rules)
3. Ability to establish and maintain rapport with participants
4. Lots of practice

2.1.1 **Understanding usability testing**

To be an effective moderator, you need a firm understanding of what usability testing is all about. You need to know the following:

What Is a Usability Test?

There are several types of usability test, but the most common is a diagnostic test that takes place during product development. Such tests have these elements:

- The primary focus is on usability.
- Actual or potential users participate.
- Participants attempt tasks with the product.
- Participants usually think aloud as they work with the product.
- Moderators observe and record behavior and comments.
- Moderators diagnose usability problems and recommend improvements.

A diagnostic usability test is *not* a market research study, a quality assurance (QA) test, or a beta test.

- the purpose of the test that you are moderating. The purpose determines how much and when to interact with participants.
- how usability testing differs from other evaluation methods. The emphasis here is on obtaining valid data from a small sample of typical users, and letting the users speak.
- how tests are designed. For example, design might be important in presenting the product and related tasks in the context of their intended use.
- how tests are planned. Every task has a purpose and the order of tasks may be important to the test's validity. Sometimes it's important to probe for an understanding of the concept behind a user interface component. Stopping a task before the participant has completed it may be required in some tests but not in others.
- how and why data is being collected. The emphasis may be on the quantitative measures or the qualitative measures, depending on the test goals.

Fundamental knowledge of usability testing is vital to understanding the material in this book, so if you need more information, we suggest that you read a book on the basics of usability tests (e.g., Dumas & Redish, 1999; Rubin, 1994) or take one of the many courses available.

2.1.2 **The basics of interacting**

The ways that practitioners interact with test participants has a huge effect on test results. Moderators (or their team mates) interact with participants throughout the testing process, including

Why Conduct a Usability Test?

- To evaluate the usability of a prototype
- To uncover problems before a product release
- To choose the best among early design concepts
- To determine the baseline of usability to measure progress
- To retest a modified design
- To compare two or more competing products

- selecting them for the study.
- greeting them when they arrive.
- providing an introduction to the test.
- guiding them as they are performing tasks.
- interacting during the post-test debriefing.
- thanking them and giving them their compensation.

2.1.3 **The ability to establish and maintain a rapport**

Our golden rules express the importance of establishing and maintaining a connection with participants. Some people seem to naturally have the "right" personality for moderating, but others don't take to it so easily. This means that some people who are new to testing will need to work harder at establishing rapport than others will. Even so, we have observed that many people in the usability profession are naturally "people people," so once they have some experience, it's easy for them to establish an appropriate rapport with test participants.

2.1.4 **Lots of practice**

Once you know the basics, becoming a great moderator only takes time and practice. Over time, you will gain enough experience and confidence to handle almost any situation calmly and effectively. (We say *almost* any situation because even after twenty years, we still encounter tests that challenge our ability to stay calm!) Practicing will help you get over the initial nervousness that all new moderators feel.

2.2 **ROLES OF A MODERATOR**

One of the things that makes moderating test sessions different from, say, conducting interviews or moderating focus groups is that you must simultaneously fill multiple roles. As a moderator, you must be unbiased and neutral with regard to the product, be in control of the session, and be open to and approachable by participants. Getting this mixture of neutrality and attachment right is the challenge of moderating effectively. One of our favorite quotes comes from Carolyn Snyder's book on paper prototyping: "A test facilitator is like a duck—serene on the surface, but paddling like heck underneath" (Snyder, 2003).

When moderating, your primary roles will be:

- The Gracious Host
- The Leader
- The Neutral Observer

2.2.1 **The Gracious Host**

Your job as a moderator does not begin and end in the test room. You are responsible for making participants feel welcome from the moment they arrive to the moment they leave. This means attending to their physical comfort, ensuring that the session goes smoothly, and ensuring that they have a positive experience overall.

A Gracious Host makes it a point to do the following:

- Greet the participant warmly, either in person or over the phone in a remote session.

- Offer to take his or her coat, bags, and such.
- Offer refreshments (at least a glass of water or a cup of coffee).
- Offer them breaks.
- Accommodate them in every way possible.

The last item is particularly important. The first thing to do is ask if participants need any adjustments to the test setup. For example, some participants are left-handed, so you may need to move the mouse to the left side of the computer or sit on their right side so that they can write more comfortably. Similarly, many participants wear bifocals and may prefer the computer screen to be closer. If participants have hearing, vision, mobility, or dexterity impairments, accommodating their needs may involve special transportation, assistive devices, or adjustments to the test room setup. Chapter 10 discusses the special accommodations needed for people with hearing, vision, mobility, or dexterity impairments in detail.

2.2.2 **The Leader**

One of our golden rules for participant interaction is "Respect the participants as experts, but remain in charge." This rule indicates that participants expect us to know what we're doing and expect us to take the lead. After all, we have asked them to participate in the study, and we are in control of how the session goes. Therefore, your second role is that of leader. A Leader must do the following:

- Control the pacing of the session so that it moves smoothly.
- Project a sense of confidence in the testing process.
- Take charge when software bugs appear or the product crashes.
- Know what to do when participants need encouragement, prompts, or assistance.

We discuss numerous aspects of these responsibilities, such as hiding uncertainty, projecting confidence, and suppressing negative emotions, in chapter 3.

2.2.3 **The Neutral Observer**

While participants are attempting to complete tasks with the product, your primary role becomes that of a Neutral Observer. At this point, your main goal is to collect accurate data while being a Gracious Host and remaining in charge. For example, a Neutral Observer does the following:

- Lets the participants speak.
- Asks unbiased questions.
- Neither encourages nor discourages emotional comments.
- Avoids defending the product design.

You need to be unbiased and objective and keep interactions to a minimum while providing support and encouragement to the participant when needed. Juggling these roles is a challenge we will address in detail.

2.2.4 **Other possible roles**

Although we are concentrating mainly on diagnostic and summative tests in this book, you should be aware that different roles may be required in other types of specialized usability testing. For example, in a test of a paper prototype, Snyder (2003) characterizes the roles of the moderator as Flight Attendant (who safeguards the well-being of participants), Sportscaster (who asks questions and talks to maximize information flow to observers), and Scientist (who maintains the integrity of the data).

2.2.5 **When roles change**

It's a good idea to explain to participants about your changing roles in the pretest instructions. It's also good to avoid changing roles abruptly. For example, if you start off being very chatty and then stop talking without warning, it could be disorienting to the participants. When switching from Gracious Host to Neutral Observer, tell participants that you are going to try *not* to talk to them, except to ask clarifying questions.

2.2.6 **When roles conflict**

Of course, your roles may conflict at times. In particular, your roles as Gracious Host and Neutral Observer might conflict when you ask participants to struggle with a difficult product or task. They may become frustrated or upset and blame themselves, and as Gracious Host you may want to remove their discomfort but as Neutral Observer you may need to let them continue. We have more to say about this conflict in chapter 3 under golden rules 2 and 3.

2.3 **TESTING LOCATIONS**

Usability tests can be conducted in a variety of locations using a number of techniques for capturing data. Table 2.1 provides an overview of advantages and disadvantages you may want to consider because they affect how you interact with participants.

When choosing the best location for your test, consider these factors:

- The amount of testing you plan to do
- The number of people who will observe
- The locations of participants and observers
- The type of product you are testing
- Your budget and schedule
- The types of interactions you need to have with participants

2.4 **TEST PREPARATION**

In this section we review the basic tasks involved in preparing for a usability test. Preparation is key to being relaxed and in control when the participants arrive.

2.4.1 **Planning ahead**

A successful moderator is well prepared. Although every test is different, preparation typically includes the following steps:

- Decide on the goals of the test.
- Design the test protocol.
- Determine the number and demographics of participants.
- Determine the location of the test.
- Schedule test dates and times.
- Prepare a screener to recruit participants (or hire an agency).
- Determine the tasks users will perform.
- Prepare the test script.
- Arrange for participants' compensation.
- Determine the equipment needed (hardware, software, video recording, etc.).
- Prepare other materials needed (consent forms, task cards, receipts for compensation, etc.).
- Set up the testing area or lab.
- Perform sound and video recording checks.
- Conduct a practice (pilot) test.
- Refine the test script and materials if necessary.

Table 2.1 Common Testing Locations

Location	Advantages	Disadvantages
Participant's environment	■ Participants don't have to come to you. ■ It may be easier to recruit participants. ■ Participants may feel more comfortable and relaxed. ■ Participants have access to their tools and resources. ■ Participants who use assistive technologies can use their personal setups, which are usually customized.	■ You have less control over the situation and environment. ■ Participants may experience distractions and interruptions. ■ You have to bring all your test materials and recording equipment with you. ■ Logistically it is a bit more difficult for you (getting directions, travel time, etc.). ■ Your interactions with participants may disrupt other people who are trying to work. ■ It's difficult (or impossible) to accommodate observers.
Internal usability lab	■ You have complete control over the environment, which makes moderating easier. ■ You can create a comfortable viewing environment for observers. ■ Observers and participants are away from distractions and interruptions. ■ Having a dedicated space makes iterative testing easier and quicker. ■ It's particularly convenient when participants are internal to the development organization. ■ It's more convenient for the development team to come and observe. ■ It's convenient for testing physical products and devices that require a lot of setup.	■ It requires a dedicated space. ■ The initial setup for a fully equipped lab can be expensive. ■ Testing can't be anonymous because you are bringing users to your own building. This could introduce bias.
Commercial market research facility or usability lab	■ You have complete control over the environment, which makes moderating easier. ■ Independent usability labs are set up specifically for usability testing. ■ Independent labs have trained usability staff that can help you with moderating, note-taking, data analysis, and reporting. ■ There is usually a comfortable viewing environment for observers. ■ Observers and participants are away from distractions and interruptions. ■ It provides anonymity to help reduce bias. ■ It is convenient for testing physical products and devices that require a lot of setup. ■ Many facilities can recruit participants for you, ensure that they arrive, provide hosting services, and handle compensation.	■ Outsourcing costs money. On the other hand, you can obtain a specific quote for the cost. ■ Market research facilities are usually not set up for usability testing, so you may have to subcontract video recording or other technologies. ■ You need to plan ahead and bring the product and all your testing materials with you. ■ You are not in your usual environment, so you may be more apprehensive during the first few sessions.

Continued

Table 2.1 Common Testing Locations   *cont . . .*

Location	Advantages	Disadvantages
Remotely via the Internet	■ It increases the participant pool because geography is no longer an issue. ■ All the advantages of testing in the user's environment apply. ■ Some new moderators find it easier to moderate remotely because they are not in the spotlight. ■ There is some evidence to suggest that participants may be more honest in their opinions when tested remotely.	■ As a moderator, you have less control over the environment. ■ Participants may experience distractions and interruptions. ■ If you do not use cameras, you are not able to see participants' faces or body language.

We like to use checklists to ensure that we don't forget any of these steps. Obviously, each step requires further explanation if you have never moderated a test. For more information on test preparation, refer to Dumas and Redish (1999) or Rubin (1994).

2.4.2 **Planning for many tasks**

For many new moderators, the hardest thing about running a test is multitasking all of the things that have to be done in addition to actually interacting with the participants. For example, you may be responsible for

- setting up the product or software for each participant.
- taking notes during the test.
- recording task times or other quantitative data.
- watching the elapsed time, and deciding what to do if a participant will not finish.
- handling technical difficulties with the product.
- running the recording equipment or software.
- interacting with clients and other observers or support staff.

Our advice is to be as practiced and prepared in these things as possible, so you can give your full attention to the participants rather than worrying whether the DVD is still recording. Some tips:

- If you have to perform setup tasks with the product or software between test sessions, practice before the test, and make sure you have a checklist of everything you have to do. Also, remember to leave enough time between sessions.

- Prepare a data collection sheet or file (if typing notes) ahead of time, and have a good labeling and numbering scheme so that the notes don't get mixed up.
- Know well in advance the types of measures you will be collecting, and have a consistent way to record them.
- Know the product and the tasks well enough that you can anticipate problems participants might have, whether technical or usability related.
- Decide ahead of time what you will do if one or more participants are not able to finish all tasks. For example, prioritize so that the last tasks are least important and can be omitted, or establish a time limit for each task.
- If you are running recording equipment, practice setting it up several times before the test, and play it back. There have been times when we have discovered too late that we had video with no audio or audio with no video! This can be a big problem if you need to create a highlight tape or if important stakeholders are unable to attend and want to watch the recordings later.

2.4.3 **Understanding the domain and product**

It is important to understand as much as possible about the product you're testing. Test results can be compromised because moderators who do well under normal circumstances are testing (for whatever reason) something that they do not fully understand; their moderating and often note-taking suffers.

At the same time, keep in mind that the participants often have much more expertise than you do in the subject of the test. Be careful not to act like a "know-it-all." Let the participants teach you a little about their field—after all, you recruited them because they are experts in a certain domain and you want to hear their perspectives.

2.5 **JUMP-STARTING YOUR MODERATING SKILLS**

In this section we provide guidance for new moderators who want to or need to get started quickly.

2.5.1 **Six things you can do first**

The following suggestions will help you jump-start your moderating skills and form a solid foundation for your moderating career:

1. Read and understand the ten golden rules we present in chapters 3 and 4.
2. Watch the videos that accompany this book because we show both good and poor practices.
3. Observe a few usability test sessions, either in person or pre-recorded.
4. Use checklists to stay organized.
5. Practice and have a colleague critique you (preferably one with moderating experience).
6. Attend a short course on conducting usability tests.

Of course, once you get started, it is equally important to monitor yourself and continually improve and adapt your skills to new situations.

2.5.2 **The big challenges**

We have had the opportunity to train many new moderators over the past twenty years. New moderators commonly have trouble with two situations: starting the session and managing a struggling participant.

Starting the session

Many new moderators experience performance anxiety, especially when they are being observed by their managers or important clients. The experience is a lot like the first few minutes of a speech. To minimize your anxiety at the beginning of a session, you can take several actions:

- Make sure you are well prepared.
- Arrive early, 10 to 15 minutes before the test, and make sure the equipment and the materials are ready.
- Memorize the first two or three sentences of the pretest briefing while imagining that you are looking at the participant.
- Take several deep breaths just before you go to meet the participant and continue to breathe deeply.

When participants struggle

When participants seem to be having difficulty, the new moderator's intuition screams, "I need to help this person so they won't feel bad about themselves!" This emotion may show in the moderator's body language and in the need to talk with the participant. It takes practice to be silent and watch. Most inexperienced moderators say too much and say things they wouldn't if they had more experience. New moderators need practice to stay in the Neutral Observer role during stressful periods. They also need to be assured that every moderator has difficulty with this and that saying things you wished you hadn't is a normal part of moderating.

■ **FIGURE 2.1** A moderator takes a minute to relax before a session.

INTERVIEW WITH A NEW MODERATOR

1. What moments caused you the most stress?

I think the anticipation right before my first session was the most stressful. I was just so worried that I was going to forget something during the briefing, or say something leading while the participant worked on the tasks.

After having gone through it a couple of times now, the part that still stresses me out is when the participant asks a direct question, such as, "Was that the right way to do that?" I worry about what to say so as not to annoy them but not to give anything away. I think it is an art to be able to answer a question with another question aimed at uncovering what you (the moderator) want to know from the participant.

2. Did you find it more or less stressful to run remote sessions versus in the lab? What's the difference, if any?

As a rookie, the idea of moderating in general was pretty nerve-wracking. I felt more stress about moderating a face-to-face session in the lab as compared to a remote session. One reason for this might be that I had observed someone else successfully moderate remote sessions, so I was less intimidated. Plus, as a new moderator, I find myself constantly worrying about things like talking too much, my tone, and wanting to help the participant. During the face-to-face sessions I am trying to process what's happening, but I'm also coaching myself to remain silent, to allow them to struggle, to remind them to think aloud if they are not.

Moderating a remote session reduced some of the mental burden for me because I was not influenced by the participant's body language or my own perceptions of their reactions to the product. I had to rely only on what I saw on screen, and what they said, to make inferences. I think this allowed me to concentrate on the task at hand and worry less about saying or doing something to unduly influence the participant. I didn't have to coach myself as much.

I also felt less stressed about moderating the remote session because I thought that if I was unsure of something, I could put the participant on hold, almost like a "time out" without them knowing, and ask one of the senior staff what to do.

I guess the technical difficulties that can arise during a remote session can add to the stress, but I think it was mitigated somewhat by the fact that I had experimented with the remote testing software and learned how to use it in advance of the tests.

Golden rules 1 through 5

In this chapter, we present the first five golden rules for interacting with test participants. We divided the rules into two chapters to avoid having a chapter that was twice as long as the others. The choice of which five rules to put into this chapter was not arbitrary; we believe that these five are the most important, so we call them our core rules.

Each rule describes an important piece of the foundation on which effective moderating stands. We expect that other moderators may have a somewhat different set of rules or may disagree with some of the details of ours. But the rules provide a stake in the ground to begin the discussion about best practices for moderating a test.

The five rules in this chapter are the broadest in scope. They lay the foundation for effective moderating. These core rules are:

1. Decide how to interact based on the purpose of the test.
2. Respect the participants' rights.
3. Remember your responsibility to future users.
4. Respect the participants as experts, but remain in charge.
5. Be professional, which includes being genuine.

The rules are in rough order of priority. Certainly the first three are the most important because, to some extent, they influence the rest of the rules in this and the next chapter.

In chapter 4, we cover the five remaining rules:

6. Let the participants speak!
7. Remember that your intuition can hurt and help you.
8. Be unbiased.
9. Don't give away information inadvertently.
10. Watch yourself to keep sharp.

3.1 **RULE 1: DECIDE HOW TO INTERACT BASED ON THE PURPOSE OF THE TEST**

We put this rule first because it has the biggest impact on moderating and it influences many of the subsequent rules. The purpose of a test is influenced by at least three factors:

1. The type of test
2. The product's stage of development
3. Your relationship with developers

3.1.1 **The type of test**

As we discussed in chapter 1, there are several types of usability test. In the most common type of test, the purpose is *diagnosis*—that is, uncovering design issues, both positive and negative. (These are sometimes referred to as *formative* tests.) Your goal is to understand the participants' mental model, that is, the way *users* think the user interface is structured, or to explore the conceptual model underlying the user interface, that is, the way *designers* have structured the interface. Some tests may explore more than one user–interface concept. In a diagnostic test, therefore, there is much interaction with participants. You can be more interactive in a diagnostic test because the goal is to find issues and explore alternatives, not to measure performance. You probe and explore to understand what is happening and why because that helps with the diagnosing of problems and suggesting solutions.

In addition to diagnosis, there are two other common reasons for conducting a test:

- To measure or benchmark the usability of a product
- To compare the usability of two or more products

In these types of test, there is less interaction between a moderator and participants. Benchmark tests, one kind of *summative* test, require a more quantitative approach to measurement because their purpose is to measure usability or to provide a baseline against which to measure the success of future products or versions. Benchmark tests are weighted more toward clean measurement than diagnosis, which typically means fewer interactions. For example, if you want to measure task times, you should avoid interacting with participants too much or letting them talk about issues that are only tangentially related to the task. Even the act of thinking out loud can sometimes lengthen task times. In addition, assisted and unassisted task success rates can be affected by the number and types of interaction with participants.

In a comparison test, the test design avoids bias for or against any one product. You need to avoid bias in the way you give feedback, provide encouragement, and provide assistance. For example, when participants experience a series of task failures; you may want to provide encouragement to keep them motivated. Or when they have a successful task following a number of failures, you may want to praise them. But you need to be much more careful about how you do this in comparison tests so as not to favor one product. We will have more to say about this issue in chapter 4 when we discuss rule 8, "Be unbiased."

It is also very difficult to give assistance fairly in a comparison test. Providing assistance can influence task success and change participants' subjective perception of a product's usability. We will have more to say about this issue in chapter 4 when we discuss rule 9, "Don't give away information inadvertently" and in chapter 6 when we discuss when and how to provide assistance.

3.1.2 **The product's stage of development**

The purpose of a test tends to be correlated with the stage of product development. Even within the category of diagnostic tests, there are differences in procedures and measures as the product design becomes more mature. Typically, in earlier stages of design, you need to interact with participants more than in later stages. Early stages may test prototypes of varying degrees of fidelity with the goal of exploring design alternatives. When testing prototypes there often are bugs to work around, or if the prototype is static, the moderator needs to explain what would have happened if it were interactive. In contrast, later diagnostic tests have less moderator intervention to make the product work or to explain what might happen.

Paper prototyping is a special case. With paper prototypes, the moderator is much more active in asking questions and sometimes draws the developers into the conversation with the participant. We will have more to say about moderating these types of sessions in chapter 9. Snyder (2003) provides the best discussion about moderating a test of a paper prototype.

3.1.3 **Relationship with developers**

Your relationship with the design and development team can also affect how much you interact with participants. Table 3.1 contrasts two situations. The left column describes interactions when you and the development team trust each other and have an effective working relationship. The right column describes interactions when the relationship is poor or uncertain.

In addition to the items in the table, other interaction strategies might emerge when you are uncertain whether developers will agree to address an issue. For example, when there is little trust among team members, you may feel the need to emphasize an issue by manipulating the participants to get them to vocalize a problem. This practice is not a good one, but you may be tempted to use it. For example, a participant says, "At this point, I would throw the product against the wall!" and the moderator, who heard the remark quite clearly, asks, "What did you say?" This response is likely to stimulate the participant to say more negative things about the product.

Table 3.1 Impact of Moderator's Relationship with Developers

Good relationship	Poor or uncertain relationship
When a problem is identified, the team agrees to fix it. There is no need to see it again and again with many participants.	A problem needs to be documented with many participants to "prove" its existence.
You do not need to have participants struggle for long periods with a problem.	You allow participants to struggle to motivate developers to address the problem.
You are free to provide assistance to move participants along because developers will address the problem.	You wait longer to provide assistance or do not provide it to avoid making the product look more usable than it is.
You are free to explore interesting and potentially fruitful diversions.	You stick to the test script to ensure that issues are addressed and documented.
You are freer to abbreviate or skip tasks that are not uncovering usability issues.	You focus on collecting measures of failure that will motivate developers to make changes.

We are not saying that you should not ask for more information when it will throw light on a usability issue. But asking for more information is different from manipulating the participant to repeat negative comments.

Finally, during the post-test interview, you may say, "Were there other problems that you saw?" This leading question may be intended to get participants to discuss and vocalize an issue that is already clear to you from their performance. We will have more to say about this when we discuss rule 8, "Be unbiased," in the next chapter.

3.2 **RULE 2: PROTECT PARTICIPANTS' RIGHTS**

Organizations and individuals that conduct tests have certain ethical responsibilities, one of which is to be aware of and protect the rights of participants. Whether you conduct formal usability testing in a laboratory, ad hoc testing with participants in a cubicle, or remote sessions, you must obtain the informed consent of all participants and protect their confidentiality. Next we discuss the key issues regarding ethical treatment of test participants.

3.2.1 **Compensation**

You should provide reasonable compensation for their time and effort. In fact, you should compensate participants even in the following circumstances.

- They stop participation during the pretest briefing. For example, participants may be uncomfortable signing the form without their company's approval.
- They stop participation at any later time for any reason.
- You have to stop the test for some reason (e.g., the product crashes).

If a participant arrives so late that you cannot conduct the session, you will have to decide whether it was his or her fault or was unavoidable. If the late participant is not at fault, consider compensating him or her anyway. Because these situations do not happen often, the cost of compensating participants is low. The goodwill you provide is a bonus and may benefit your organization.

3.2.2 **Informed consent**

Every testing organization should have an adequate "informed consent" form and every test session should start with both an explanation of the form and adequate time for the participants to read and sign it. *We believe that the most important responsibility you have is to obtain informed consent from participants.* See chapter 5 for a detailed discussion of the form's content. See chapter 8 for an explanation of how to obtain informed consent in a remote test.

We know of one company that does not provide informed consent for some of its participants. That company enters an agreement with companies that are its customers. As part of that agreement, the customer's company agrees to provide informed consent to any of its employees who are going to be participants in a test. They enter this agreement so that after it is signed, their employees don't have to sign additional forms. We applaud the intent but believe that this practice is unethical if any participants do not receive the information necessary to give their informed consent for a test session.

If you work for an educational or research institution in the United States, you probably have a group called the Institutional Review Board (IRB) that oversees the conduct of studies involving human participants. IRBs are important because they educate researchers about ethical responsibilities, ensure that researchers comply with guidelines, and follow up with researchers during and after their studies. Some IRBs require that you submit test protocols, recruiting screeners, questionnaires, and other materials ahead of time for their approval.

If you have any reason to believe that a participant might have an issue with informed consent or nondisclosure, send the forms to him or her ahead of time, if possible, to look them over. This practice can save you and the participant an embarrassing moment.

3.2.3 **Confidentiality**

It is your responsibility to safeguard the identities of test participants and to avoid mapping their identities to their data. You (or the test director) should be the only person who can match names to data. For example, participants' full names should not be used on forms or labels. On our test schedules (the sheets that say who is coming at what time along with their demographics), we typically refer to participants by their first names and last initials so that when they arrive, we can call them by their first names. However, during data analysis and reporting, we refer to them simply by participant number.

You must restrict the use of the data to the purpose described in the consent form (for example, the data must be used only for the purpose of improving the product, not for marketing or advertising). You should be especially careful about the use and distribution of highlight video clips. If you are a consultant, this is of particular concern because even though you explain to your clients that the video clips should not be used for any other purpose, you don't have control over their distribution once they are in the client's hands. Similarly, the unedited recordings of test sessions typically become your client's property. Be sure to tell your clients that they are responsible for the confidentiality of the data, and don't provide the last names or other identifying data about participants.

Confidentiality is of particular concern when the participants are employees of the organization developing the product you are testing. If the participants' manager or senior executives plan to observe the test session, you have a special responsibility to tell participants *before* the session, so they are not faced with the pressure of deciding whether to participate while being observed by their supervisors. In tests such as these, you may want to avoid creating video highlight clips or you may insist that only you or member of the team can show a tape.

An issue related to confidentiality is informing participants about *their* responsibility to keep the knowledge they gained about the product confidential. When this is an issue, there is usually a confidentiality form (often called a nondisclosure agreement, or NDA, form) for participants to sign. Some organizations combine the consent and the confidentiality forms. We think that it's best to keep them separate. Confidentiality forms are prepared by lawyers. They are dense and require close reading to comprehend. Most participants skim and sign them, but some spend a long time reading them, which is their right. Just as with the consent form, it's preferable if you address confidentiality in plain English and give the participant some words to say if anyone asks what they were doing. For example, "This is our confidentiality form. It says that what you see today is confidential and that you can't talk about the product, its usability, or its performance to anyone. If you need to tell someone at home or work about what you were doing here today, just tell them you were evaluating the usability of a new product. If they ask for more information, just say you signed a form that says you cannot give any details."

3.2.4 **Balance of purpose and risk**

As usability professionals, we must ensure that the purpose of our studies justifies the risks to participants. Fortunately, very few usability testing situations involve physical risk. Even so, we know that there is some level of emotional stress involved. Although we tell participants that we are not testing *them*—we are testing the product— many participants still blame themselves when they struggle or fail. And although we should never pressure or coerce participants, we sometimes ask them to continue with a task even when they say they're stuck.

We believe that the justification for putting participants under some stress comes from the fact that future users will be spared a bad experience. We discuss this further in rule 3, "Remember your responsibility to future users." As you will see, even that responsibility has its limits.

3.2.5 **Priorities**

Sometimes the test moderator has to make decisions on the spot to handle a situation in which the participant is very, even unduly, stressed. In terms of priorities, the order of protection should be the participant first, the organization second, and the integrity of the data third. After ensuring the well-being of the participant, you have a responsibility to your organization to collect the test data, as well as to maintain positive relationships with customers and the public. In addition, you do not want potential (or actual) users of the product to leave a test feeling mistreated because that would reflect badly on the organization.

Although you were hired to and want to gather valid data, never forget your responsibilities to the participants and the organization in the pursuit of it. It is far easier to throw out the data from a test session and run another one than it is to repair damage to the participants' well-being or the organization's reputation.

3.3 **RULE 3: REMEMBER YOUR RESPONSIBILITY TO FUTURE USERS**

This rule is seldom discussed but provides the rationale for many of the toughest decisions a moderator will have to make. It is the other side of the ethics coin. Although you must respect and protect the rights of participants, it is not necessary to protect them from all unpleasantness, which in itself might provide very valuable data. There are times when you have to let test participants struggle and fail. Allowing these sometimes unpleasant situations is your responsibility to the *future* users of the product.

3.3.1 **Letting participants struggle**

Every time a usability problem is fixed, future users of the product have been spared from it. When something serious is fixed, you have saved possibly hundreds or thousands of users from a worse experience than the first participants in the test have faced. Consequently, the benefits of finding serious problems and fixing them are enormous. A few test participants who struggle and thereby motivate the development team to improve usability can save future frustrating experiences.

Of course, it would be ideal if participants did not have to struggle and fail. Usually when a participant struggles for more than a few minutes and/or fails a task, the cause is a usability problem. But sometimes we're not sure, so we need to test the product with more participants, or we need more or different data to understand what is causing the problem. Furthermore, sometimes we can't be sure that developers will address a problem without substantial data to document it (particularly if your relationship with developers is uncertain; see Table 3.1). Developers may need to see every participant fail a task to be motivated to make a change. This situation is frustrating for everyone—the participants, you, and the developers.

You sometimes can avoid that situation. If developers' attitudes about users and usability are going to change, sometimes they need to see the participants struggling in real time. Watching sessions live is much different from watching them on a recording. For reasons we don't fully understand, people tend to sympathize and project their emotions onto others much more readily in a live situation. In fact, when observers watch a test session live, they are often more anxious and frustrated than the participants are. This experience has made usability advocates of many developers and managers.

In addition, there are some types of test during which the moderator can't intervene to prevent or stop failure. For example, in a baseline or comparison test to collect quantitative measurements, you may not be able to avoid or truncate a task that almost everyone fails.

3.3.2 **Conflicting responsibilities**

There comes a point when two responsibilities conflict—to the participants and to potential users. It's important to keep your responsibility to future users in mind as participants struggle and fail, so long as you are not in conflict with your ethical responsibilities to the participant.

It is ethical to allow participants to have negative experiences until it begins to evoke strong feelings or becomes counterproductive. At that point, your ethical responsibility to the participant takes precedence and you must intervene. Remember that the ethical treatment of participants supersedes other priorities. The comfort of future users never comes before the breakdown of emotions or self-esteem by the participant. In chapter 6, we discuss how to deal with failure in greater detail. That will be the most difficult situation you face as a moderator.

3.4 **RULE 4: RESPECT THE PARTICIPANTS AS EXPERTS, BUT REMAIN IN CHARGE**

One of the reasons we tell participants that they can't make a mistake is because they are experts at what they do, whether it is to structure a database or shop online. They are evaluating the product for us. We need to treat them accordingly. However, they also expect us to know what we're doing and expect us to be in charge. After all, we have asked them to participate in the study, and we are in control of how the session goes. This expectation is why your role as Leader is important.

3.4.1 **Controlling logistics and pacing**

Part of making participants feel at ease is showing them that you are comfortable and at ease. This means projecting confidence and staying on top of the pacing and logistics of the session.

One of the reasons we strongly advocate running a practice test session (often called a "pilot" session) before the test is to work out uncertainties, test the wording of tasks, and fine-tune the pacing. That way you are confident and in control when you run the first real session. However, this makes moderating a practice session even more challenging than moderating the remaining sessions!

Should you tell pilot participants that they are the practice participants? In general, we avoid telling them this because it could bias them and/or make them less confident that we know what we're doing. If the script is straightforward and we are testing an existing product, then we are pretty certain that the practice test will run smoothly, and we don't tell the participant that he or she is the first one. On the other hand, if the script is complex or we're testing an unstable prototype, we can expect some difficulties, so we may tell the test participant that this is the first session and ask him or her to "bear with us."

3.4.2 **Stopping unacceptable behavior**

Although it is rare, occasionally a participant acts inappropriately. It may consist of a touch, some words, or even a look. As a moderator, you do not have to let these behaviors pass. When a participant is doing something unacceptable, you need to address it directly and as forcefully as you see fit. This situation is an exception to our admonitions about not showing negative feelings to participants. You should never feel that you are being abused in any way, nor should other participants. If necessary, stop the session, and make sure that the participant is never asked to come in again. It is also good practice to let your colleagues know what happened and to save any recordings of the session.

3.4.3 **Dealing with uncertainty**

Sometimes the participant takes an action and the product reacts in a way you didn't expect. Avoid reacting too quickly. Pay attention and see what happens next— many times the participant will recover and move on. If you're not sure what's happening, and you think it will affect the session, stop and ask the development team. Participants will accept this. If you are being honest about the situation, it will show in your voice and body language. You might say, "I'm not sure whether that is a bug or not" or "Hmmm, I'm not sure why that happened. Let me find out."

It's perfectly acceptable to excuse yourself from the testing room to consult the development team. Simply tell the participant you need to check with your colleagues about what is happening. It will seem more professional than if you continue to appear concerned or confused. These days it is also possible to be in contact with a developer through a chat application or instant messaging when you need to ask a question.

If the product or prototype is particularly buggy, you may need the developer to come into the test room and make some changes. If the developer has to work on the prototype for more than about 5 minutes, we suggest that you take a break with the participant and leave the room (avoid talking about the test, though). If the trouble-shooting takes more than 15 minutes, we feel it is unfair to ask the participant to wait, so consider ending the session.

3.4.4 **Projecting authority**

An important aspect of your Leader role is to maintain a calm, professional, and matter-of-fact demeanor. You need to be practiced enough to avoid showing nervousness. Over time, you will become less nervous and more confident. If you are new to moderating, practice! Here are some ideas.

- Practice using coworkers as test participants. You can run a mock test of anything just to gain experience in giving instructions, taking notes, thinking on the fly, and conducting an interview. Taking notes while listening can be a very difficult skill to master.
- Run a practice test session and have a more experienced moderator observe and give you feedback.
- Run a practice session and record yourself. When you watch the video, we guarantee you will see things you can improve upon. (Even after twenty years, we see things we can do better!)
- Watch someone else moderate and note what he or she does well and not so well.

Inexperienced moderators are often most nervous at the start of the session, but then they relax as the session goes on. Since the start of the session is when we give participants their instructions, review their rights, and explain what will happen, it is usually helpful to have a checklist or even read the test instructions verbatim. This dispels some nervousness and ensures that you don't forget important points.

Of course, the best way to avoid nervousness, even if you are a seasoned moderator, is to know the product and the script very well. In some cases, we run more than one pilot test—a "prepilot" using an internal participant and a "final" pilot with an actual end user we have recruited.

A final trick is to greet the participants and establish a connection before the session starts, preferably outside the test environment. We find that once we establish a connection, we're less nervous. We talk about this initial contact in detail in chapter 5.

Of course, there isn't always time to learn much about the product or to run multiple practice sessions. The pressure to get the results of the test as fast as possible sometimes means that you're not as prepared as you would like to be. In these situations, do the best you can and do not blame yourself for what is outside your control.

A friend of ours, Mary Beth Rettger, has the best advice we know of about self-blame: She gives herself permission to do or say one stupid thing every day. When you realize that these things happen regularly, it's easier to put them behind you and move on.

3.4.5 **Dispelling negativity**

An effective moderator does not carry negative emotions into a testing situation. You may be frazzled and upset because you have just spent hours getting the product installed or had a tough commute. Your nonverbal cues can convey a sense of negativity to participants. It's difficult to fake being positive when you don't feel positive because you give off unconscious cues to your mood. In this situation, you might need to take a breath, or perhaps several, in order to relax and clear your mind of negative thoughts and focus on the task at hand. It's better to do this, even if it keeps the participant waiting a few extra minutes. One of your responsibilities is to make a connection with participants and make them feel appreciated.

■ **FIGURE 3.1** A moderator instructs visitors about observation etiquette.

3.4 6 **Managing visitors**

One aspect of being in control is instructing visitors about appropriate behavior. When visitors, such as managers or developers, come to observe a test, it is almost always a positive sign. You want them to come because it increases the likelihood that improvements to the product will be made. In this section, we discuss what you need to do when visitors come to an in-person test. In chapter 8, we will discuss managing visitors in a remote testing situation.

First, take extra steps to welcome visitors. Call or email them to remind them about the session. Have drinks and refreshments available. Give them a tour of the facility if they are unfamiliar with it and if they arrive early. Explain your roles.

Some testing teams involve visitors in the session by having them write down their observations and any issues they see. This strategy is most common when there is a good working relationship between the usability and development teams.

When visitors come to watch a usability test for the first time, they often don't know what role they should take. For example, people in marketing and management are used to demonstrating new products. They're likely to want to jump in and show users what the product offers. They sometimes know a lot about the product but are not good listeners. Consequently, you must give visitors some guidance about what you expect of them.

The visitors' main role is to watch what happens. You can facilitate observation by showing visitors the best place to watch from. Depending on the physical arrangement, the best viewing can be on a display, through a one-way mirror, or from a particular place in the test room.

Provide visitors with a copy of the test script so they can follow along, and warn them about laughing out loud. Participants may hear the laughter, even when there is soundproofing. You don't want this! Participants almost always interpret laughter as directed at them.

If there are several visitors, designate one of them to be the spokesperson. When you have a chance to speak to him or her privately, ask whether the person would like you to ask the participant a question or to tell the participant something about the product, but take it as a suggestion, not a demand. As the moderator, *you* decide what to say and when to say it. If the request is reasonable, you may follow

Suggested Rules of Conduct for Visitors

- Maintain silence. Speak softly, if necessary. Don't laugh out loud or carry on cell phone conversations.
- Do not touch the control room equipment.
- Avoid letting excess light into the observation room.
- Close doors quietly if you must enter or exit the room while the test is in progress.
- Avoid speaking about participants before or after the session . . . they might be standing behind you!
- Do not discuss confidential issues.

it, although usually with some rewording to remove bias. If you decide not to follow the request, you can say, "That's an interesting question, but I don't want to ask it right now. I will tell you why after the session."

Be firm about enforcing the rules. Some visitors will violate the rules. Visitors who come for several sessions often assume that they are not going to see or hear anything new. Consequently, their attention wavers and they start talking among themselves. When you are in the observation room with them, usually a gentle reminder works best: "Let's watch and see if she does what the others have done." But sometimes you just have to ask them diplomatically to be quiet: for example, "I can't hear what he is saying." If you are in the test room with the participant and you can hear noise from the observation area, find a way to pause the session and go into the area to let the visitors know that you can hear them.

3.5 RULE 5: BE PROFESSIONAL, WHICH INCLUDES BEING GENUINE

This is the most difficult rule to explain, but we feel that it's important enough to be a core rule. As we have said, an effective moderator establishes an emotional connection with the participant. You can follow all of the other rules but still not be effective if you are aloof or clinical. By "effective," we mean getting the most information from participants while making them as comfortable as possible. You need to be friendly, warm, and approachable to establish and maintain a connection. You have to be genuine.

How do you create and maintain a connection? Rather than try to explain this concept in abstract terms, we present a list of actions you can take and actions to avoid.

3.5.1 Dos for making and maintaining a connection

This list is based on our experience interacting with participants primarily in North America. Some items may be different in other cultures.

Greet them warmly. Participants form an impression of you quickly, in the blink of an eye. The way you first approach them is important. If you don't look directly at them or if you act distracted, participants may not connect with you. In the United States, it's a common practice to extend your hand to shake theirs, especially if you have never met them. A handshake conveys a positive feeling. And keep in mind that participants often are more nervous than you are!

Look them in the eyes. "The eyes are the mirror of the soul" is a saying that stresses the importance of eye contact in human communication. Making eye contact with an audience is one of the keys to effective public speaking. A similar principle applies to test moderation. If you constantly look at a checklist or the product, participants will also look at them and lose their connection with you.

Smile. A smile is an important part of a warm greeting. In our experience, smiling does not come naturally to some people, so they may need to remind themselves to smile when greeting participants. Watch for the reaction you get when you do smile as part of a greeting. It affects both you and the person you greet in a positive way.

Sit at the same height as participants. Whenever possible, meet people at their level rather than making them meet you at your level. If you're taller than average, lower your chair so you are not above the participant's head. Don't stand over them.

Hold a relaxed posture. Tension and uneasiness can be conveyed unconsciously by body language. Stiffness in your hands, neck, and facial muscles can be cues to inner tension. Breathing deeply and relaxing your muscles will help convey positive energy rather than negative.

Listen attentively. Look at participants when they are talking and use an occasional nod of the head to indicate that you are listening. Convey that you are interested in everything they have to say, even if you have heard the comment a hundred times. Participants should feel appreciated and that their opinions are valued.

Use "acknowledgment tokens" periodically. Acknowledgment tokens are sounds such as "ah huh" and "mm hmm" that convey the message that you are listening and comprehending what participants are saying without taking active control of the conversation. In addition, your tone of voice as you acknowledge participants can convey that you *care* about what they're saying. These sounds are especially important when you are interacting over an intercom or phone. A friend of ours reports that at one time she was conducting usability tests during the day and teaching yoga classes in the evening. A colleague of hers happened to attend both sessions and remarked that our friend used the same voice: deliberate, slow, calm, and focused.

Use their names occasionally and correctly. People respond to their names. They listen more intently and are more likely to remember what is said before or after they hear their names (Mitchell, 2007). Don't use participants' names casually. Save it for when you want them to be sure to hear what you're saying, such as when they are stressed or need encouragement. An example might be, "John, remember we talked about how you can't make a mistake here?" or "You're really being helpful, Sue. This is exactly what we need."

Very soon after greeting the participants, ask the names they prefer you use. Don't assume that a shortened form or nickname is preferable without asking. Some Michaels prefer "Michael" rather than "Mike." Many Americans are comfortable using first names from the start of a relationship, but some are not. People from many other cultures think it is rude for someone they have just met to refer to them with less than a formal name. So ask participants what they prefer and then use that name when you want to give emphasis to what you say.

Use a modulated tone of voice. Changing the loudness and pitch of your voice allows you to use your voice for emphasis and to convey a range of emotions. This is especially important when you are speaking to participants over an intercom or phone.

Speak slowly. There is a lot for the user to grasp and understand. Some moderators speak too quickly and risk losing their connection with the participant.

Adapt to the participant's interaction style. Some participants speak quickly, and others speak slowly and deliberately. Some speak loudly and others speak softly. Although it may be difficult to make major shifts in your loudness or pace without sounding artificial, you can make some adjustments to match the participant's style. But be careful about speaking too softly. When participants speak softly, you will have a tendency to match them, and this can make your conversation inaudible to observers or the recording. Remember that you are speaking to the microphone in addition to each other.

3.5.2 Don'ts for making and maintaining a connection

Again, this list is based on our experience interacting with participants primarily in North America. Some items may be different in other cultures.

Don't act distracted. Sometimes when a session begins, your mind is racing. You may have just spent a tense hour trying to get the product working or changing the test script at the last minute. When you're unfocused, you may worry about forgetting everything you want to say, so you start looking around for your materials, forms, and so on. You may start running through a checklist in your mind when you should be communicating with the participant. These behaviors convey the message that you're tense and care more about getting your procedure right (or getting the session over with!) than making a connection. If you find yourself in this state, take a break and relax. It's good to have a checklist to make sure you remember the important points, but review it before the test and use it after the test as a final check (Dumas & Redish, 1999).

Don't use an unmodulated, flat tone of voice. It's amazing how the same words can sound very different depending on the qualities of your voice. "You're giving me great feedback" can sound phony and aloof when said with a flat unemotional tone, or it can sound positive and reinforcing when said with a warm and inviting tone.

Don't exhibit nervous or repetitive habits—pen tapping, giggling, clearing your throat, and so on. Even experienced moderators fall into repetitive habits that can be annoying. These actions are almost always done unconsciously, so they're hard to detect. You can't get rid of them until you become aware of them. Make sure you periodically watch a recording of yourself moderating, or have a colleague watch a session and give you feedback.

Don't overuse acknowledgment tokens or names. Be careful about the overuse of these important techniques. Don't acknowledge every phrase or sentence that participants utter. (Have you ever seen a moderator say, "Uh huh . . . uh huh . . . uh huh," through an entire session?) Overuse turns an effective communication tool into an annoying habit. The same applies to overusing the participant's name. It can make you sound phony.

Don't rush participants or cut them off. It's tempting to do this when you have heard the same comment from many previous participants or when it's late and you want to go home—but you have to seem just as interested and engaged as you were with the first participant. Wait until he or she finishes before you speak.

Don't show annoyance with participants. There are times when you may be annoyed. They may have habits you don't like; they may remind you of someone you don't like; or they may appear to be trying to do as little work as possible. If you want to be effective as a moderator, don't show your annoyance as long as participants are acting appropriately. This is part of being a professional moderator. You can go off by yourself after the session and vent your frustration, but don't do it during the session.

Don't yawn. Yes, you may be tired. This may be your sixth test of the day. But participants will interpret yawning as a sign that you are bored with them even when it's not true. Because yawns often are involuntary, they sometimes happen. If you find yourself yawning, just calmly say, "Excuse me" and move on.

Don't touch the participant, other than a handshake. While touching someone's arm or shoulder may communicate warmth to some people, it can be upsetting to others. It may seem overly friendly or it could be misinterpreted. Unless you know the participant well, do not touch him or her during the session. You cannot predict the reaction to such contact.

3.5.3 **When you're not in the room**

It's very challenging to establish and maintain a connection when you're not in the room with the participant. When you interact with participants over an intercom or over the phone in the case of a remote test, you have only your voice to convey your feelings. It's important to speak to the participants frequently and to modulate the loudness and tone of your voice.

If you are conducting a summative test where measurement is important, you may need to minimize interaction. In this situation, you're more likely to lose the emotional connection with participants. In a diagnostic test, on the other hand, you have more freedom to be engaged. Keep in mind that talking over the intercom or phone has two purposes: a primary purpose (for example, to ask the participant to move on to the next task) and a secondary purpose, which is to maintain the connection. To achieve both purposes, *how* you communicate is just as important as *what* you communicate.

INTERVIEW WITH AN EXPERIENCED MODERATOR

Have you ever touched a participant?

I knew it would be a challenge for me when Beth and Joe explicitly said in their UPA tutorial, "Don't Touch the Participant." I'm a very tactile person. I hug and kiss people when I greet them. I let kids climb all over me. I've never been hurt by a touch and touching is part of connection and affection and warmth. I don't hug my boss or anything, but I do have coworkers with whom I'm on a hugging basis.

But I forgot this once in a test session. It happened like this:

We were using phone and Web conferencing to allow remote observation of our test, so the development team and product managers, in different states, could watch the sessions live. I had set up the phone conference on the phone next to the computer in the participant's room. Later, I would moderate and log data from the control room next door.

I had ushered in the participant and I had told him about recording, about breaks, and a little about what we would be doing. I had not mentioned the remote viewers yet. I had asked him to read and sign the informed consent, as well as the confidential disclosure agreement (CDA).

As the participant turned to read the CDA, he accidentally knocked the handset off the phone. Then he put it back, cutting off the conference call. I think I let out a little gasp or said "uh-oh" or some such thing. "What?" he asked. I said, "It's OK. We have observers on a conference call." "Did I just cut them off?" "It's OK. We'll just dial back in."

But then, because I was moderator of the conference call, I had muted all other callers. When I dialed back in, I came in muted. This would certainly make it hard for observers to hear the participant! I ended up going into the control room to find a way to tell everyone else on the call to hang up, and then I dialed in again from the participant's room.

So the participant was there, with his head in his hands over the CDA, saying "I'm so sorry!" as I came to dial back in. I needed to pass near him to get to the phone. My heart went out to him, he felt so bad. Without thinking, I patted him on the back as I got near the phone. "No, don't worry about it at all. It will be fine."

So there I was, looking at the phone, talking to the participant, when I realized that I had my left hand on his back. "Oh, no! I'm touching the participant!" I thought. I took my hand away and said nothing about it. He didn't seem to have noticed—at least he didn't say anything about it. I just went ahead, did the pretest interview, had him practice thinking aloud and all that. And then I went to the other room during the tasks. At the end of the session, we wrapped things up and I ushered him back out the front door, making sure not to touch him again except to shake his hand when I thanked him for his participation.

By the way, this was the first of a two-part test: two hours in one session and two hours in a second session. The participant did come back for the second session, and I made sure to be on my best professional behavior. I didn't touch him in the second session, except for the handshakes at the beginning and the end.

4

Golden rules 6 through 10

In this chapter we cover the five remaining golden rules, which build on the set of core rules we presented in chapter 3:

- Rule 6: Let the participants speak!
- Rule 7: Remember that your intuition can hurt and help you.
- Rule 8: Be unbiased.
- Rule 9: Don't give away information inadvertently.
- Rule 10: Watch yourself to keep sharp.

4.1 **RULE 6: LET THE PARTICIPANTS SPEAK!**

One of the most obvious things that distinguishes experienced moderators from inexperienced ones is the amount of talking they do while participants are working on tasks. The experienced moderator speaks only when necessary and watches and listens more, carefully observing and analyzing what is going on. Less experienced moderators, on the other hand, interrupt participants while they're talking or try to finish their thoughts, sometimes putting words in their mouths or giving elaborate explanations for actions they've taken.

It's important to let participants speak, rather than dropping into a conversation. One reason is that you're there to learn from them, to listen so they can teach you about their particular knowledge and experience. Also, from a practical standpoint, you're more likely to get good quotes if you are not talking over participants. Unfortunately, we've seen many a great video highlight clip ruined by the moderator's leading prompts!

During the introduction to the session, you need to tell participants how you're going to interact with them. For example, tell them you are going to try to talk as little as possible, and ask them to try to figure things out for themselves. This will prepare them for your lack of responses. We show an example of this in the videos on the book's web site.

Here are some ways to ensure that you're letting participants speak:

- Allow participants to stay in control of the communication. This is called *speakership*.
- Use interruptions appropriately.
- When in doubt, wait before speaking.
- Remember that saying nothing is still communicating.

We expand on these guidelines next.

4.1.1 **Speakership**

"Speakership" refers to who has control of a conversation. In normal conversation, people pass speakership back and forth through several mechanisms, including verbal confirmations ("What do you think?") and nonverbal cues like looking directly at the person, raising the eyebrows, and making a hand gesture. Of course, these mechanisms vary by culture. When you're moderating a test session, the participant should maintain speakership about 80% of the time (Boren & Ramey, 2000; Dumas, 1988). If you want participants to keep talking but don't want to take control of the conversation, use acknowledgment tokens (see page 34).

4.1.2 **Appropriate interruptions**

Whenever possible, let participants work on tasks without interruption. This will make the experience seem more natural because users typically interact with a product without someone continually asking, "What is happening now?" or "Tell me what you see on this screen." If participants feel comfortable and you have instructed them on how to think out loud, you shouldn't have to interrupt very often. The fact that you are asking them to think aloud can affect the participants' thought processes (see chapter 10 for a discussion of the limitations of thinking aloud for some populations). For a demonstration of coaching participants on thinking aloud, see the videos on the book's web site.

Keep in mind that this guideline should be balanced against the amount of probing you need to do. As we discussed in rule 1, the appropriate amount of interaction depends on factors such as the product's stage of development and your relationship with developers

4.1.3 **Judicious speaking**

We have found that waiting before speaking often eliminates the need to speak at all. Some moderators teach themselves to count to ten before interjecting. This is because participants often ask questions while thinking aloud that they don't really expect you to answer, such as "Hmmm, so what does *this* do?" They're just working through a problem out loud and don't need a response.

Waiting before speaking is especially important with older adults and participants who have cognitive disabilities. These populations are more likely to be distracted by interruptions and lose their concentration.

Other times, participants will ask questions that they *do* expect you to answer, but if you wait before answering, they may figure it out on their own. This can give you valuable insights as to what was confusing and how they determined the correct course of action.

Overall, the less you speak, the better, because every time you do, you risk taking control of the conversation.

Admittedly, this is difficult when you have participants who constantly say, "I need help" or "Was that right?" In this case, you'll need to remind them that you really would like them to do the best that they can without help because when they have difficulty, it's actually helping you learn how to improve the product. (We say this a lot!)

4.1.4 **Silent communication**

If the participant asks a question and you don't answer it, then he or she is free to interpret what your silence means. This is a tricky one. Some participants are timid and frequently ask for confirmation, even if you have explained that you would like them to figure things out on their own. If you say nothing when they ask you a question, they may think that you didn't hear them. Or they may feel that you are testing *them* and not the product. (We've had participants say things like, "You're not going to tell me, are you? You're going to make me guess!") The best thing to do is to remind them that you would like them to continue working and that you'll answer any questions they have at the end of the session.

Sometimes you can lighten up the situation and work toward both connecting with the participant and making him or her feel more comfortable just by smiling and saying something like, "Sorry, but I can't tell you that right now. We can talk about it after you've evaluated the whole product."

4.2 RULE 7: REMEMBER THAT YOUR INTUITION CAN HURT AND HELP YOU

Can your intuition be both helpful and hurtful? We have watched many usability professionals transition from moderating their first sessions to becoming experienced moderators. We created this rule because the transition people go through is fairly consistent.

4.2.1 Your intuition can hurt you

We believe that the majority of usability practitioners—particularly those who moderate usability sessions—enjoy interacting with people and can easily empathize with others. Because of this, they have a natural tendency toward wanting to please people and are good at making people feel comfortable in artificial testing situations.

The downside is that when you're new to moderating, your intuition will, to some extent, work against you. You will want to say too much. Our advice is to resist the temptation to explain and elaborate.

In addition, you'll have to resist the natural tendency to make the participant feel better about failure. You need to be friendly to participants without being a friend. We will talk more about this in chapter 6.

4.2.2 Your intuition can help you

With time and practice, your intuition will become helpful again. You will incorporate the rules of interacting with test participants into your skill set. Without thinking about it, your intuition will tell you what to do, and it will tell you when you violate a rule; listen to it. You will know when you've said too much or you've said something awkward or biasing. Forget it and move on.

By the way, we've never run a test session without thinking afterward that we could have done it better. There is no such thing as a perfect test session.

4.3 **RULE 8: BE UNBIASED**

One of the most critical aspects of successful moderating is to avoid biasing the participants. You can unintentionally bias participants in a number of ways:

- through a biased test script
- through biased questions
- through biased answers
- through nonverbal cues

If you bias participants, the data could be compromised. If the data is compromised, it could call into question all of the results of the study and possibly even undermine your credibility with the development team.

4.3.1 **Use an unbiased test script**

Because this book is about interacting with test participants, we're not going to discuss the details of developing an unbiased test script,

■ **FIGURE 4.1** A participant fills out a rating form.

so for basic information, look at Dumas and Redish (1999) or Rubin (1994). Suffice it to say that tasks should be worded in a neutral manner, probing questions should be unbiased, and in a comparison test, tasks should be equitable to all of the products. Additionally, you should understand the basics of designing a balanced test protocol.

4.3.2 **Use unbiased questions**

It's crucial to be unbiased when you ask participants for an opinion or a preference. Many inexperienced moderators make serious mistakes here. Ask questions in an open-ended manner rather than show your opinions or conclusions. You need to remain unbiased and open-minded to be truly neutral in your questioning.

Keep in mind that people often hear the dominant thought in a statement, which makes it difficult to avoid bias. We now know that if you ask, "Did you like that or not?" a participant is likely to hear "Did you like that?" So adding "or not" to a statement or appending the opposite adjective or adverb to a statement doesn't mean you have avoided bias. If you are going to use this phrasing, make sure you give equal weight to both answers, perhaps pausing in between, such as "Did you *like* that, or did you *not* like that?"

It's best to avoid using adjectives and adverbs. Instead, ask a more open-ended question such as "What did you think of that?"

Sometimes it's hard to recognize your own biases, especially if you're close to the design. This is where a two-person testing team can be an advantage. Ask someone on the research or development team to keep an eye on you while you're moderating and to let you know if they notice biases creeping in.

Here are some additional guidelines for unbiased questioning:

Watch out for adjectives and adverbs such as easy, hard, *and* helpful. Try to avoid using these words, and instead ask one of the neutral questions. This will encourage users to provide their own adjectives.

Be consistent from participant to participant. If you ask one participant a question, ask *all* the participants the same question. (Note: This rule only holds for planned questions, not spontaneous ones.) If there's an important issue that you want to probe, then plan a question about it and ask all participants. (See section 6.3, When and how to probe.) However, you don't have to repeat spontaneous questions unless the answer was particularly insight-ful. In that case, you might want to write it down and ask future participants.

Be consistent from task to task. If you ask a question such as "How confident are you that you completed that task?" then do it for every task; otherwise, you might draw unwanted attention to that one task.

4.3.3 **Keep answers unbiased**

It is also important to give unbiased answers when participants ask *you* questions. For example, if two participants ask, "Did I do that right?" and you tell one that she did and decline to answer the second, then the first participant has information that the other doesn't. This information could be useful for future tasks, making them easier and affecting the results. A helpful technique is to turn the question around. For example, participants will often ask for an affirmation, such as "Did I do that right?" Instead of replying "Yes" or "No," you might say, "Did you think that was the way to do that task?" If a participant asks, "Why is it asking me for my Username?" you might say, "Why do you think it's asking for that?"

Another important technique is to give similar feedback for both negative and positive comments about the product. If a participant says, "Wow! That's a really great feature," you should respond neu-trally and say, "OK, thank you for the comment." If the participant later says, "I hate that feature," you should respond in the same way, "OK, thank you for the comment."

When participants continue to ask questions or seek affirmation, it's possible that they do not understand your Neutral Observer role, so you may want to refer back to the instructions, reminding them that part of your role is not to lead or bias them. Otherwise, you may end up repeating your response and you could sound evasive.

4.3.4 **Watch nonverbal cues**

One of the hardest things to avoid is biasing participants through your body language or facial expressions. In fact, many practitioners believe it is unavoidable, so they choose to moderate from outside the room. (We discuss this further in chapter 9.) However, assuming that you plan to moderate from inside the room, here are some suggestions:

- Maintain a neutral posture; don't lean toward or away from the participant.
- Maintain fairly neutral facial expressions.
- Learn to sit still. If you fidget while participants are working, they might think you're bored or that they are doing something wrong.
- Give the same nonverbal feedback (e.g., a nod of the head) for both positive and negative comments about the product.

If you take notes, take a lot of notes. If you don't, participants may be self-conscious when they hear you write something down. If they don't hear you write anything, they may think the feedback they gave was not important. Worse, they may think you're only recording instances when they're successful or making mistakes, and this could influence their behavior. The alternative is to take very few notes and rely on another person to take notes during the session, and you can view the recordings to take notes later.

INTERVIEW WITH AN EXPERIENCED MODERATOR

How do you take notes during a session?

"Great point about taking lots of notes. I often take notes on a laptop. If I keep my hands on the keyboard, I don't have to make a large gesture to take the note, and I can type a lot of the time. It also lets me break eye contact gracefully— especially when I'm working with professionals who also spend a lot of time on a computer—so I can subtly "pull back" to let them work on their own."

4.4 **RULE 9: DON'T GIVE AWAY INFORMATION INADVERTENTLY**

Providing participants with assistance can help you uncover more usability problems. But sometimes the assistance can influence the test in a negative way.

4.4.1 **Giving an assist**

An "assist" is an intervention on your part to move participants toward task completion. (We discuss how to give assists in more detail in chapter 6.) Basically, you give assists in the following situations.

- Participants are stuck, but you believe there is more to be learned if they continue.
- Participants have unknowingly gone too far down the wrong path and you need to bring them back on track.
- You want participants to move on to the next task because you already know about the problem they are having and/or time is limited.

Avoid giving participants information that they would not get on their own, especially if it will help them do a subsequent task. If you assist too much, they may end the session feeling that the product was easier to use than it was and give it a higher subjective rating.

Unfortunately, sometimes you can't avoid giving an assist that may help the participant complete a future task, but you should at least know when you are doing it and consider its impact when you analyze the data.

4.4.2 **Explaining the designer's intent or being defensive**

It is very difficult to moderate test sessions when you are designing the product because you are too familiar with it. You know all about the goals of the design, the underlying technology constraints, the legacy systems that had to be accommodated, the trade-offs that were made along the way, the politics involved, plans for future enhancements, et cetera, et cetera. It is similar to trying to proofread your own writing. This is one reason that development teams often ask people from outside the team to moderate a usability test.

Note: During the introduction to a test, we usually tell participants that we did not design the product and that we are neutral third parties, so they should be open and honest with all of their opinions.

If you find yourself in the position of moderating a test of a product that you have designed, you need to avoid being defensive. Participants will stop being honest very quickly if they feel that you are defending the existing design rather than being open to feedback and suggestions for change.

Regardless of whether you have been involved in the design, it is important that you avoid explaining the design team's intent. For example, after participants struggle with a task because they do not understand the concept behind a design element, you might be tempted to explain the concept to participants. These explanations may help participants and you feel better, but they are not good practice. They can make you sound defensive and/or they suggest that you are committed to the design.

There is one exception to this practice, however. We sometimes allow participants to struggle with a task in order to gather information about what is confusing. Afterward, we explain the intent of the design just enough so that participants understand it, then ask for feedback about how they think it could be made clearer.

4.4.3 **Recording all suggestions**

In most situations, it is not necessary to go into detail about future product plans. It is better to write down all suggestions without comment or critique, and thank participants for their ideas. Test sessions should be used to *evaluate* products, not to *design* future products. Additionally, users cannot know all of the other factors that will influence the future, nor is it their job to know. Don't give release dates or feature commitments that you cannot guarantee. Sales representatives become very angry when participants call them up asking why they can't have the features they saw in a test session.

4.5 **RULE 10: WATCH YOURSELF TO KEEP SHARP**

We all develop bad habits; it's not a failure. As we mentioned, even after moderating tests for many years, we still make mistakes. We sometimes catch ourselves asking a question in a less-than-neutral way or fail to keep the session focused. Here are some suggestions to help stay sharp and avoid developing bad habits.

We all hate to watch ourselves; it's human nature. However, it is important to record yourself periodically and review your moderating habits. We often do this while we're creating highlight clips since we have to review the recordings anyway.

It helps to practice with a colleague and give each other feedback, especially if you are new to moderating. You can even act out different participant personality types (the difficult participant, the participant who forgets to think aloud, the "cold fish," etc.) to make it more interesting (Beauregard, 2005).

Look for annoying gestures, repeating words, facial expressions. (We knew someone once who said "basically" in every paragraph until someone pointed it out.)

Moderating usability tests well takes skill and practice, but it also takes vigilance. Keeping yourself sharp will benefit both you and the participants.

5

Initial contacts

You never get a second chance to make a good first impression, right? In this chapter we discuss the critical time *before* the test starts, when you have the chance to make a great first impression, establish rapport with participants, and put them at ease. You can avoid many problems during the test by taking the time to set the tone, establish the "ground rules" (including how you will interact with participants), and set their expectations for what will happen during the session.

In this chapter we focus on interaction with participants at three points of contact:

- during recruiting
- when they arrive
- during the pretest briefing

5.1 **RECRUITING**

We have learned through experience that how you recruit participants, and the information you give them when doing so, makes a big difference. If you do this poorly, or provide incomplete information, you may get the wrong participants or they may arrive with incorrect expectations. Actually, even when you've done a good job, some participants still arrive with incorrect perceptions and expectations. (It's amazing how many of our participants still arrive thinking they are participating in a focus group!)

In this section we talk about choosing a contact method, explaining what the test is about, screening candidates for eligibility, and confirming their appointments.

5.1.1 **Contacting participants**

There are a number of ways to recruit participants for a usability test; the best method depends on the nature of the test and the types of people you're looking for. Many practitioners hire third-party recruiting agencies and pay them (per hour or per person) to locate, screen, and schedule test participants. Other practitioners do their own recruiting, building up a database of potential participants over time. Sometimes the client (the group sponsoring the test) provides a list of potential candidates from which to recruit, particularly if current customers are desirable participants. Be aware, however, that sometimes this is a big help and sometimes it makes the recruiting harder because you're so constrained or because the information is out of date.

Sometimes clients want to do the recruiting themselves to save money, but we advise against it and suggest you avoid this if possible. The pitfalls include that the clients may

- underestimate the time recruiting takes from their already busy schedules.
- get a "convenience" sample, bypassing the screening process, which could skew the results.
- inaccurately convey to participants the nature of the test (especially if the clients are unfamiliar with usability testing themselves).
- make promises that you cannot keep (e.g., inappropriate incentives, scheduling sessions when you are not available).
- be unable to recruit enough participants and ask you to take over the recruiting at the last minute.

Regardless of who does the recruiting, you need to decide how to reach out to the target population. The most common methods are advertising via the Internet, phone, email, or a combination of these. Quite a lot has been written about recruiting, so if you need further guidance, consult books such as *A Practical Guide to Usability Testing* by Dumas and Redish (1999), *Understanding Your Users* by Courage and Baxter (2006), or posted articles such as "233 Tips and Tricks for Recruiting Users as Participants in Usability Studies" by the Nielsen Norman Group (2003).

For advice on recruiting participants from special populations, such as children or persons with disabilities, see chapter 10.

5.1.2 **Advertising**

If you're recruiting via an advertisement (either printed or posted on the Internet) or via email, then keep it very short and think carefully about what you want to say. You will need to clearly state the following information:

- who your ideal participant is
- when and where the study will be held
- what you will require of participants
- what you will give them in return (e.g., an appreciation gift)
- whom to contact if they are interested

Figure 5.1 is a typical Internet posting and Figure 5.2 shows an email sent to recruit employees for a study of the company's intranet. In these examples, we are asking interested candidates to

---

Do you use an online pharmacy?

Seeking 25- to 64-year-old women and men for web site study

Bentley College is recruiting individuals who use online pharmacies to participate in a web site usability test at its campus in Waltham, MA, on April 18, 19, or 20. Participants in this 2-hour study will receive $100 cash for taking part.

If you are interested in participating, please email [*recruiting agency*] with the following information:

- Names of online pharmacies you use
- Your name
- Age
- Gender
- Occupation
- Marital status
- Daytime phone number

Because of the overwhelming response we sometimes get from online postings such as this, we may not be able to contact everyone. We are looking for specific user profiles, so be sure to include all of the information requested here.

■ **FIGURE 5.1** Sample Internet recruiting posting.

Participate in a usability study and earn a $20 gift certificate!

We are recruiting [*company*] employees to participate in a usability test of your company's intranet. We need 18 people—mix of clinicians, managers, and administrative staff—to participate in the study. During the week of August 7–11, you must be able to give us 2 hours of your time and come to the Smith building.

You will receive a $20 online gift certificate for taking part in the usability test. If you are interested in participating, please reply to this message with the following information:

• Name
• Job title
• Phone number
• Email address

When your message of interest is received, we will be in touch to ask you some additional questions to see if you qualify for the study. Thank you for helping with this important project!

■ **FIGURE 5.2** A recruiting email sent to company employees.

email the recruiter, but you can also provide a link to an online questionnaire that they fill out. Sometimes that is easier for potential candidates.

5.1.3 **Explaining the test**

Your initial contact with potential participants sets the tone for all subsequent interactions. To be sure that all the necessary information is exchanged and that unnecessary information does not sidetrack the conversation, we recommend that callers use a script to screen participants. Figure 5.3 shows part of a script for telephone recruiting.

The script should include the following information:

■ who you are and the company you're representing
■ a clear statement that you are not selling anything; you are looking for people to participate in a one-on-one product study
■ the dates, times, and location of the study

Recruiting College Students and Their Parents for Web Site Study

I am calling today because we are looking for people who might be interested in participating in a usability project. The Bentley Design and Usability Center is conducting a study for a company that provides financial services to college students. This business wants to make its web site more user friendly and is looking for feedback from current college students and their parents. As a token of appreciation, each participant will receive $75 in cash.

We are specifically looking for pairs consisting of an undergraduate college student and one of his or her parents. Is this something you [*and your parent/son/daughter*] would be interested in?

If the answer is yes, continue with this:

Let me tell you a little more about the study. The interview will take about an hour and a half and will be held at Bentley College in Waltham, MA. Your participation will be scheduled for a specific time between September 7 and September 14. During the interview, we'll ask you to try using the company's web site and to provide feedback on it. You will be in a room with a Bentley moderator who will lead you through the session. Your [*parent/child*] may be in the same room with you or may be in a separate room.

If you have a few minutes, I'd like to ask you some questions to determine whether your background matches the study.

Do you have time right now to answer a few questions?

Your answers are confidential and will be used only to determine whether you're eligible to participate in this study.

If no, arrange a time to call back.

■ **FIGURE 5.3** A telephone recruiting script.

- any incentive you are offering
- assurance that their identity will be protected and the data you collect will remain confidential
- disclosure that the session will be recorded
- necessity to sign a nondisclosure agreement (if appropriate)
- a question asking if he or she is interested, and if so, does he or she have time to answer a few screening questions

INTERVIEW WITH AN EXPERIENCED MODERATOR

Do you have any tips for moderators about recruiting via email?

I find that it's important to tell candidates when to reply to the ad and to have a cut-off date. I also go out of my way, on a separate line, to tell them that we may not respond to them. With hard-to-find groups, like physicians and highly paid professionals, you may want to send a "rejection" letter to those not selected and tell them they might be eligible for the next study. Then you can contact them in the future.

Another thing I do is instruct the recipient to reply to the email and cut and paste the text; that way I can include some direct questions and they don't have to retype them or risk missing some of the categories. I have also had great success with sending an email introducing the study topic and including a link to a survey. In the survey I ask screening questions or even branch some questions like in a phone survey.

Stating clearly that you would like to record the session is key. You don't want someone coming in and objecting to that because you'll have to pay them anyway and you lose the ability to record (and maybe even observe) the session. This happened to me recently when I let the client do the recruiting.

When people respond, I ask them to list all of the times they are available, and then I choose the ones that fit the time slots best (for example, I've done this with physicians when I had a limited set of candidates to choose from). Some candidates are more flexible than others. Then I take the list of candidates and select the best candidates rather than ones that just happened to fit into a time slot.

How you explain what the test is about, and how you ask the demographic questions, will set the tone for potential candidates. You should use short, clear sentences, and a friendly voice, and express enthusiasm for the study. You should be open to answering any questions they have (except perhaps the name of the company or product, if that needs to remain confidential). It's important to convey to them that they will be judging the product and providing valuable feedback and that you will not be testing them or their abilities. Sometimes we say, "There are no right or wrong answers; we are just very interested in your opinion."

There may be some cases when you need to ask some sensitive questions. For example, over the years we have had to recruit people with specific illnesses, certain attitudes toward reproductive rights, and people who have a child with limitations. In these cases, you need to give extra thought to how you craft your posting and script, as well as the questions you need to ask.

5.1.4 **Screening candidates**

Use a list of questions to ensure that candidates meet the criteria you need to represent the target population accurately. We recommend starting with the most important questions that will determine their eligibility (e.g., Have you purchased a car in the last six months?) because there is no point in asking other demographic questions if they don't meet the main criteria. For each question, make a note to the recruiter as to which responses are acceptable and which are not. In addition to the eligibility questions, you may want to ask some "for information only" questions; for example, while recruiting people who have used a travel reservations web site, you might want to know what other sites they have used to purchase services or products. But keep these information-only questions to a minimum. You can always ask additional questions in a pretest questionnaire.

It is a good idea to test the screening questions ahead of time to make sure they are clear and that you haven't overlooked possible conflicts. The logic can be tricky, especially if you're screening people to put them into several subgroups. For example, if they answer that they are current users of the product, you may place them in the "Customers" subgroup and then ask them additional relevant questions, or you might ask questions that do not suit noncustomers. Figure 5.4 shows part of a typical screening questionnaire.

When asking demographic questions, be sensitive about the wording. For example, some people prefer not to give their exact age, education level, or income level, and some prefer not to disclose their ethnicity. Don't ask these questions unless they are an important part of the recruiting criteria. If such information is part of what the client is trying to find out, then a range usually suffices and yields valuable data. For example, if the client is trying to find out if middle-aged people buy a lot of music online, the question could ask the range of the respondent's age; that is, the respondent chooses from 18–25, 26–44, 45–60, 61 and older). Of course, the ranges can be wider or narrower depending on the study's purpose.

Obviously, you will have to disqualify many candidates because they don't match the recruiting criteria. This can be difficult to do, especially if the candidate really wants to participate in the study. The way to handle this depends on your specific circumstances and the nature

Screening Questions

1. What is the gender of the student? (*Recruit a mix*)
   _____ Male
   _____ Female

2. What is the age of the student? (*Must be between 18 and 21 years*) _____

3. Is the student enrolled in a degree or certificate program at least half time?
   _____ Yes   (*CONTINUE*)
   _____ NO   (*DEFER*)

4. What year is the student entering this fall (2005–2006 school year)?
   ❏ Freshman ⎤
   ❏ Sophomore ⎦ (*Recruit about 75%*)
   ❏ Junior ⎤
   ❏ Senior ⎦ (*Recruit about 25%*)

5. What college does/will the student attend? _____
   (*Recruit a majority [approximately 70%] of public/state schools, if possible.*)

6. Is a parent involved in looking at financial options for the student (e.g., grants, loans, scholarships, work study)?
   (*Reword if speaking to parent.*)
   _____ Yes   (*CONTINUE*)
   _____ No   (*HOLD: Can use them as an individual student but not as a pair.*)

7. Would you and your parents/child be likely to look for this kind of information online?
   _____ Yes   (*CONTINUE*)
   _____ No   (*DEFER*)

8. What types of financial options are you using or considering to fund your education?
   _____ grants or scholarships
   _____ federal student loans
   _____ parental loans
   _____ money earned while at school (work study or job)
   _____ money from savings
   _____ other _____

9. What are the occupations of both parents?
   _____
   (*Exclude bankers, members of the financial industry, market researchers, and Web designers.*)

10. Would the parent or student consider conducting any of the following transactions online? (*Check all that apply.*)
    ❏ Making a purchase
    ❏ Filling out a college application
    ❏ Filling out a job application
    ❏ Applying for a credit card
    ❏ Applying for a loan (*DEFER, if the respondent says "No"*)
    ❏ Applying for a mortgage
    ❏ Setting up an online bank account

11. What is the annual household income? (*for information only*)
    ❏ Under $50,000
    ❏ $50,000 to $99,999
    ❏ $100,000 to $150,000
    ❏ More than $150,000

Great! It looks like you have the right background for the study. (*Proceed to gather contact information and schedule participants.*)

■ **FIGURE 5.4** Sample screening questions.

of the target population (for instance, whether they are members of the general public or a subset of your most important customers); the key is to be as polite and gracious as you can be.

We typically go through the entire screening, and then tell the person the truth: either that his or her background doesn't exactly match what is needed for this study or that we already have enough people who match his or her background. Depending on the situation, you may want to defer, rather than disqualify, a candidate if he or she is close to the ideal but not an exact match. You may also tell a candidate that although it appears you have recruited enough participants of matching background, you would like to keep him or her as potential replacement should someone else cancel. Finally, you can assure an eager candidate who can't participate in this particular study that he or she might be perfect for a future study and that you will keep the information on file.

5.1.5 **Confirming appointments**

Once you have selected participants and scheduled them for their appointments, it's critical to follow up with them. Sometimes one follow-up phone call, letter, or email is sufficient; other times making two contacts is better. For example, you may want to follow up immediately with a letter or email and then call or email again right before the test.

When you follow up, use a friendly tone and stress how important participants are to your study. Remind them that it will be a one-on-one session so you are counting on their arrival. Clearly indicate whom they should call if they need to cancel and whom to call if they get lost on the way. Ask them to arrive 10 to 15 minutes before the session, which helps ensure they arrive on time. You might also state that you'll reimburse their parking or transportation costs (if appropriate) and that light refreshments will be available. The goal is to reduce all possible barriers to getting to the right place at the right time.

5.2 **WHEN PARTICIPANTS ARRIVE**

You might recall that chapter 2 provides a list of common tasks to prepare for a usability test. By the time the first participant arrives, you should have everything ready and organized so that you can give your full attention to the participant. If you are not prepared and emotionally grounded, you will have a harder time making a connection with participants and making them feel comfortable.

5.2.1 **Greeting participants**

Treat participants as people who are giving *you* something valuable. After all, they have chosen to come and help you design a better product. Some tips:

- Look the person in the eyes, smile, exchange names, thank him or her, and offer a handshake.
- Get a sense for the participant's mood. Is she nervous, tired, angry (e.g., had a difficult ride to the site)? If so, deal with it now.

Confirm the participant's identity both to keep the data straight and to be sure he or she didn't switch with someone else. Watch for the "professional participant," who is someone who exaggerates or lies about qualifications to obtain the incentive. Consider asking for identification if you are suspicious (many market research facilities do this).

■ **FIGURE 5.5** A moderator greets a participant with a handshake and a smile.

5.2.2 **Creating comfort**

Remember that one of your roles is that of Gracious Host. You are responsible for making participants feel welcome from the moment they arrive to the moment they leave. This means attending to their physical comfort, ensuring that the session goes smoothly, and ensuring they have a positive experience overall.

5.2.3 **Obtaining informed consent**

Every testing organization should have an adequate informed consent form and every test session should start with both an explanation of the form and adequate time for participants to read and sign it. This form serves the legal purposes of documenting that participants were informed, the specific information they received, and their agreement to participate. In addition, some people simply take in information better if it is written than if it is spoken to them.

The consent form should address the following:

- what will happen during the session
- the participant's right to stop the session at any time and still receive the incentive
- the right to take a break
- disclosure that you will be recording/videotaping (if applicable)
- disclosure that others may be observing or listening (if that is applicable)
- the way(s) data and recordings will be used
- assurance of confidentiality and the steps taken to protect participants' identities

Any test session conducted without informed consent is unethical and anyone who conducts tests without a proper consent form violates the code of ethics of professional organizations such as the American Psychological Association, the Human Factors and Ergonomics Society, and the Usability Professionals' Association.

We often give participants the consent form when they arrive in the waiting area so that they have time to read it without someone looking over their shoulders. In fact, since the participant is giving permission to be videotaped, it is not proper to start recording until they have signed the form, which is another reason to take care of it before bringing the participant to the test room. Even though they may have signed the form while waiting, you should still review it with them to ensure that they understood the information.

Figure 5.6 shows a typical informed consent form. You can find other sample forms in Snyder (2003), Dumas and Redish (1999), and Rubin (1994).

Understanding Your Participation in This Study

Purpose

Bentley College and (*vendor*) are asking you to participate in a study of the _____ web site and its ease of use. By participating in this study, you will help us improve the site's design, making it easier to use.

Procedure

In this study, we will ask you to perform a series of tasks using the _____ web site. Afterward, we will ask about your impressions of the site. The session will take approximately 60 minutes. We will use the information you give us, along with information from other people, to make recommendations for improving the web site.

Recording

We will be recording where you go on the web site (video) and the comments you make (audio). The recording will be seen by only the Bentley team that analyzes the data and members of the _____ web site development team. After the data analysis is done, we will send the tapes to the client's design team. The videos will be used to improve the usability of the web site, not for any other purpose.

Confidentiality

Your name will not be identified with the data in any way. In addition, only Bentley College employees who are working on the project will have access to the data we collect.

Risks

There are no foreseeable risks associated with this study.

Breaks

If you need a break at any time, just let us know.

Withdrawal

Your participation in this study is completely voluntary. You may withdraw from it at any time without penalty.

Questions

If you have questions, you may ask them now or at any time during the study. If you have questions after the study, you may call us at 781-891-2500 or email us at usability@bentley.edu.

By signing this form: You are indicating that you agree to the terms stated here and that you give Bentley College permission to use your voice, verbal statements, and videotaped image for the purpose of evaluating and improving the company's web site.

Signature: _____

Printed name: _____

Date: _____

■ **FIGURE 5.6** Sample informed consent form.

WHAT THE RESEARCH SAYS

Making informed consent forms usable

This study varied the format and wording of informed consent forms (Waters, Carswell, Stephens, & Selwitz, 2001). The original form was the typical informed consent form used at their university. It was written in the first person but contained no headings, much jargon, and stilted text, for example, "I have been given the chance to ask questions about the study, and any questions have been answered to my satisfaction." A second form, the revised form, used a FAQ format and more plain English, and used the second person to explain issues. For example,

WHAT IF YOU HAVE QUESTIONS?

Please feel free to ask any questions that might come to mind now. Later, if you have questions . . ."

A third form, a hybrid, contained headings that were not questions but did not use the second person to explain issues, for example, "ASKING QUESTIONS AND RECEIVING MORE DETAILED INFORMATION."

The authors inserted two inaccuracies in the three forms, one stating it would take 15 hours rather than 1.5 hours for the study and one about its impact indicating that participants might experience temporary numbness in their hands. The authors wanted to see if the participants would spontaneously comment on the inaccuracies as they read the form.

The authors gave each form to 10 students who had volunteered for a real study. After the study, they were given a test to see how much of the consent form they remembered. Their performance showed that participants who received the revised form remembered the most information and were more likely to comment on the inaccuracies. When shown all of the forms and asked for their preference, the participants preferred the hybrid form because they did not like the FAQ format. The authors' final format replaced the questions from the FAQ with the headings from the hybrid form but kept the second-person wording from the revised form.

The authors conclude the report with a few guidelines gleaned from their study:

1. Shift from legal first-person tenses to more conversational, second-person formats.
2. Identify legal, technical, and scientific words and replace with plain English.
3. Use headings to provide cues to the contents of statements.
4. Present information orally as well as on the form.

Waters, S., Creswell, M., Stephens, E., & Selwitz, A. (2001). Research ethics meets usability testing. *Ergonomics in Design*, Spring, 14–20.

5.3 **THE PRETEST BRIEFING**

Now you are finally ready to start the actual test session! In a local test, this means bringing the participant to the testing area, if you haven't already. Before you do, however, make sure that you are ready, that all equipment is present and functional, and that any observers are all settled.

5.3.1 **Preparing yourself**

The first few minutes of a test session can be the most difficult for new moderators because it's when they are most nervous. They feel it is the beginning of a "performance." They often stumble over their words and/or look extensively at their checklists or notes. Their minds may go blank momentarily. They lose eye contact and their body language and voices say, "I am new at this" to observers and participants. To minimize performance anxiety during the first few minutes of a session, there are several things you can do:

- Make sure you understand the purpose of the test and of each task.
- Arrive early, 15 minutes before the participant, and make sure the equipment and the materials are ready.
- Consider memorizing the first two or three sentences of the briefing while imagining that you are looking at the participant.
- Take several deep breaths just before you go to meet the participant and continue to breathe deeply.
- If you are new, practice, practice, practice the beginning of a session.
- Remember, if you are not relaxed, neither are the participants or the observers.

5.3.2 **Preparing participants**

We like to keep this part of the session to about 10 minutes at the most, but there is a lot to cover.

To start the briefing, tell the participant about the purpose of the test and exactly what you and he or she will be doing. The way you interact during this briefing is extremely important because it sets the tone for moderating the test. You need to convey with your manner and pace that these instructions are important, not just something to "get through." Avoid rushing or looking distracted.

You can avoid later problems by empowering the participants. Tell them that they are helping *future* users by being in the test and they can't make a mistake. Any problems they have with the product actually help because they show where the product can be improved. Remember that it's common for people to blame themselves if they can't complete a task with a product (more on this in chapter 6).

5.3.3 **Using a script or checklist**

Whether you're an inexperienced moderator or an old hand, we recommend following a script or checklist to make sure that you cover everything in the pretest briefing (see Figure 5.7 for an example). This will help you to stay focused, reduce anxiety, and ensure that all participants receive the same instructions, which is important to the validity of the data.

In the accompanying videos on this book's web site, we show examples of two pretest briefings. In Video 2, the moderator reads a script word for word. In Video 1, the moderator follows a checklist but ad-libs the words. You will see a difference. A moderator using a script covers all the points in the given order but has less eye contact with participants and the interaction can seem stilted and impersonal. On the other hand, following a checklist is more natural and seems more relaxed but is less consistent. Use the technique that works best for you.

You will occasionally forget to include some part of the briefing. This is where a checklist helps. When you get to the end of the briefing, look at the checklist to make sure you have covered it all. If not, just calmly add the missing information. For example, "By the way, I may stop a task before it is completed. If that happens, it's because I have learned all I need from that task and I want to make sure we get through most of the tasks."

Some usability professionals create a participant-friendly version of the checklist that they display on a computer screen during the introduction. They find that it eases the tension for both the moderator and the participant.

Note that the checklist asks participants to turn off their mobile communication devices (cell phones, pagers, etc.) to avoid interruptions. In some cases, the participant can't do this (for example, a child or babysitter must be able to reach them), but in most cases participants are willing to be out of touch for a short time.

Checklist for Pretest Briefing

❑ Say "Welcome" and introduce yourself.
  ■ I did not design the product, so you won't hurt my feelings with any comments you make.
❑ Describe parts of the session: how long it will last; you can take breaks at any time.
❑ Review Informed Consent
  ■ We will be recording the session, so others may review it later.
  ■ Recordings are used for research purposes only.
  ■ Other people may be watching.
  ■ Your name will not be connected with any data collected.
  ■ You have the right to stop at any time without penalty.
❑ Most important: It is the software we are testing, not you. Any difficulties you may have are because it wasn't designed in a way that makes sense to you.
❑ My role:
  ■ Neutral Observer who will be taking notes.
  ■ Clarification of the tasks themselves; otherwise, I will remain silent.
❑ Your role:
  ■ Be yourself, have fun—you can't do anything wrong!
  ■ Be candid—you are helping others shape the product at an early stage.
❑ The tasks:
  ■ Read each one aloud.
  ■ Ask for clarification if needed.
  ■ Try to complete them as if you were doing this at home.
    • Spend as little or as much time as you normally would doing similar tasks.
  ■ Let me know when you have completed each task or gone as far as you can.
  ■ It is OK if you cannot complete each task; there may not be enough time to do every task.
  ■ *Repeat:* It is the software we are testing, not you.
❑ Ask participants to think aloud.
  ■ Describe your steps: What are you looking for? Give a "narration" and so on.
  ■ Your comments are very important—we are interested in both what you like and what you don't like about the product.
  ■ Demonstrate think-aloud. [*This can be done later, before the tasks, if you prefer.*]
❑ Are there any questions before we begin?

■ **FIGURE 5.7** Sample checklist for pretest briefing.

5.3.4 **Practicing the think-aloud technique**

During most usability tests (but not all), we ask participants to think aloud while they work. For many participants this comes naturally and isn't a problem. Others benefit from examples of what we mean by "thinking aloud." You can do this by

- explaining what you want them to do.
- giving examples.
- demonstrating.
- having them practice.
- a combination of these techniques.

If you have time, we recommend that you do all of the preceding, which are shown in Videos 1 and 2 on the web site. You can do the think-aloud training during the pretest briefing or right before participants try the first task, whichever makes most sense in your situation.

There has been some discussion among usability practitioners about the best way to demonstrate the think-aloud technique when testing a web site. Initially in our lab, we demonstrated using a stapler, pen, or some other common physical object. Later, we felt that this might be too far removed from a web site and too abstract, so we started demonstrating using an unrelated web site. Unfortunately, we found that if we used a web site, participants would mimic the kinds of comments we had made when they were looking at the web site being tested. For example, if we had said, "I like this menu across the top, but I don't understand this option here," then participants would comment on the menus when they saw the web site we were testing. This problem caused us to go back to the stapler method.

In reality, you may not have time for the demonstration and/or practicing, or your company may have a different procedure. We've conducted and watched many sessions in which thinking-aloud instructions are given in different ways and have not noticed substantial differences in how the participants subsequently perform. Our data is anecdotal, but most participants catch on to the think-aloud process quite quickly and only a few participants need one or two reminders.

As far as we are aware, there is no research that looks at how different training methods affect performance, but the accompanying "What the research says" documents the inconsistencies of moderators both within and among companies. There might be a better way to instruct participants that we have not discovered yet, or it may make little difference how we do it.

5.3.5 **Confirming that participants are ready**

Before moving on, we always take a minute to ask participants if they have any other questions before we begin. We also ask if they need to use the restroom or if they would like a glass of water. Often, they decline a refreshment when they first arrive but change their minds once they feel more comfortable.

5.4 **TRANSITIONING TO THE TASKS**

Following the pretest briefing, you are ready to start collecting data. This usually means starting the participants on the tasks or giving them a pretest questionnaire and then starting them on the tasks. Since this is the first time you have asked them for information (after they have listened to you talking for so long!), it's time to switch to your role of Neutral Observer. This means remaining friendly and interested but neutral and unbiased with regard to the product and the participant's reaction to it.

5.4.1 **Starting the tasks**

Transition to the tasks by telling participants you would like them to start evaluating the product. You may want to tell them how many tasks they will attempt to complete if you think it will help with the pacing. On the other hand, if there are thirty tasks, it may be too daunting to tell participants up front. You may just want to say, "I have a number of tasks here that I would like you to work on. If we don't complete them all, that's OK." In either case, telling participants that they may not get to every task sets expectations and reduces self-blame if they only complete a portion.

Depending on your protocol, you can hand them the first task card and ask them to read it or you can read the task out loud for the participant. Sometimes we do both—we read the task aloud to participants, ask if they have any questions, and then place the task card where participants can refer to it if necessary.

There may be times when participants don't understand the task as it is written. Try to explain the task using different words, but be careful not to give away information that would help them complete the task.

5.4.2 **Conducting a pretest interview**

When possible, we like to take a minute or two to conduct a pretest interview for several reasons:

- It helps relax participants.
- It gets them talking about themselves, showing right away that they have important information and opinions.

- It gets some of their background information on the video recording.
- It provides some background information for the observers.

Often we just ask participants some simple demographic questions that repeat what was asked on the recruiting screener. We then ask a little about how they currently use the product or similar products. One advantage of conducting this interview is to allow observers or people watching the recording to hear participants' qualifications. This practice avoids doubts about whether the participants are qualified if they begin to fail too many tasks.

Other times, we use a more elaborate set of questions designed to gauge their pretest opinions and expectations of the product, which we then repeat at the end of the test to see whether their opinions have changed.

In the next chapter we will discuss the details of interacting with participants while they're working on tasks.

WHAT THE RESEARCH SAYS

How consistent are moderators?

Only one study observed test moderators while they worked (Boren & Ramey, 2000). This was more of a field study than research. Ted Boren, as part of his master's thesis, supervised by Judy Ramey, visited two established usability labs. He observed nine moderators at the two sites. All but one of the moderators held advanced degrees in psychology or technical communication, and all but one had at least four years' experience conducting usability tests. Many of these moderators were regarded by their peers as among the most methodologically rigorous at their organization.

Word-for-word transcripts of the interactions were created and the moderators were interviewed after their sessions.

The results of the observations were disappointing in that there was little consistency in some basic test practices.

- Seven of the nine moderators neither modeled nor provided training in thinking aloud. One moderator trained the participant and one moderator provided modeling.
- Reminders to keep talking ranged from short ("OK") to long ("Don't forget to tell me what you're thinking"), from directive to

nondirective, from personal to impersonal, and everywhere in between.

- The average disparity between the quickest prompt to talk and the longest delay without prompting was about 16 seconds; in other words, in the typical test session, a usability professional might prompt as quickly as 5 seconds in one instance but go as long as 21 seconds before prompting in another instance. Since the median delay before prompting was only 10 seconds, this 16-second disparity is huge in proportion to the total time participants were silent.

- Moderators inconsistently intervened to probe a particular area of the software, to help a participant who was "stuck," to request clarification of a comment, to clarify task instructions for a participant, to help participants get around software problems, and so on.

In summary, this field study showed that even moderators within the same organization do not follow the same rules when they give instructions in thinking aloud and when they interact with participants. We do not know whether these inconsistencies affect the quality of tests.

Boren, M., & Ramey, J. (2000). "Thinking aloud: Reconciling theory and practice." *IEEE Transactions on Professional Communication, 43*(3), 261–278.

INTERVIEW WITH A NEW MODERATOR

What was it like moderating your first few sessions?

At first, the moderating situation felt really unnatural. When I first started moderating, the briefing seemed endless, and I felt foolish repeating a lot of the stuff the participant just read in the consent form. But, I quickly began to realize that the participants don't know what to expect from the session either, and it probably feels even stranger to them, and they are in fact looking for me to lead them through it.

One thing that is really interesting to me is that each participant challenges you in a different way. I think at first most moderators are afraid of getting a participant who won't talk enough, which can be hard to manage. But I've also had people who talked too much, and it's difficult to deal with that too, because you want to be polite and give them a chance to give you feedback, while at the same time you want to get through all the tasks. I also had someone who seemed impatient to get through stuff, and I started feeling pressured to rush through the briefing and questions I had for them. So you're likely to learn something new from each session you run.

Chapter **6**

Interacting during the session

This chapter covers the issues that you will spend most of your time dealing with during the 60 to 90 minutes when participants are working on tasks. We focus on common situations that occur most of the time, not unusual events that rarely occur. (See Loring & Patel, 2001, for a discussion of the uncommon situations.) An effective moderator handles these typical events with no difficulty.

This chapter contains a set of "good practice" guidelines for the everyday situations:

- How much to interact
- Keeping them talking
- Probing for more information
- Providing encouragement
- Dealing with failure
- Providing assistance

6.1 **INTERACTING FOR A REASON**

We have talked about how much to interact in previous chapters. In chapter 1 we said that you talk more in a diagnostic test than in a summative test and when a product is at an early stage of development rather than near completion. In chapter 4 we made the point that the participant—not the moderator—should be talking most of the time and we introduced the topic of providing assistance without giving away information. This chapter has a more detailed discussion about these and similar issues.

Moderators have their own styles of interacting. During diagnostic tests, effective moderators fall at both ends of the activity scale, which

71

means that some interact a lot and some very little. As we have said, there is no research on whether the *amount* of interaction matters to the validity of testing.

When a moderator talks, he or she influences the results of the test in some way. Sometimes the intervention makes a test session more productive. Other times, it can compromise the validity of the data. Probes and questions should further the objectives of the test, such as by clarifying what's happening or revealing additional information.

6.2 KEEPING THEM TALKING

In most usability tests (but not all), participants are asked to think aloud as they work on tasks. Specifically, we ask participants to

- verbalize their experiences as they're interacting with the product.
- describe their expectations regarding various features of the product.
- share personal preferences and any other comments they may have.

Using a think-aloud protocol is valuable because of the rich qualitative data it produces. It helps identify areas of the product that could be improved because it documents such aspects as users' goals and expectations, areas of confusion, and unproductive paths.

One thing you learn quickly is that thinking aloud comes easily to most people but not to a small minority. Also, even a talkative participant may suddenly become quiet while working on a particular task, which is when you will need to intervene to get him or her talking again.

As we discussed in chapter 5, during the initial briefing you should instruct participants on how to think aloud. Despite your instructions, some participants will stop thinking aloud. The most common reasons are that they simply forget and that they are concentrating on a task that's taking all of their cognitive capacity.

6.2.1 Prompting as a reminder

When participants are working but not talking, it may be time for a reminder. You will need to decide how long to wait before prompting them. As the research shows, and as we have observed by watching many moderators, there is a wide variation in how long moderators wait before prompting (Boren & Ramey, 2000). There is no general rule about how long to wait before prompting because it depends on the participant and the situation. With experience, you will get a good feel for the trade-off between keeping them talking and interrupting their train of thought. (We tend to wait longer than many moderators before prompting.)

Most participants need only a gentle reminder to get them talking again. Some common prompts include:

- So . . .?
- So, you're thinking . . .?
- What are you seeing here?
- Tell me what is happening.

Our preference is for simple reminders. A "So?" spoken with a rising inflection usually works. Avoid asking questions like "Why did you do that?" because it can imply that the participant took a wrong action, or it might make him or her change the strategy for completing the task.

6.2.2 **Prompting the silent ones**

Occasionally, test participants are just very quiet, for whatever reason, and no amount of prompting will make them think aloud. In that case, you need to watch their actions more carefully and prompt them with questions in order to understand what they are thinking and trying to do. These people can be difficult to interact with because they are so reluctant to speak. It takes concentration to keep working with them and not show annoyance. But keep in mind that it is not your fault that they won't talk.

When you encounter a participant who is silent and your reminders haven't worked, there is one more tactic you can try. Interpret the silence not as silence but as moving slowly and cautiously. You might say, "It's sometimes helpful to consider what you want to say," or "It's important to really think about what you are going to say before you speak." These statements sound paradoxical, but you have nothing to lose if the participant hasn't been able to think aloud.

6.3 **WHEN AND HOW TO PROBE**

A probe is an intervention by a moderator that asks participants for additional information or clarification. Whenever one of the objectives of a test is to gather diagnostic information about the strengths and weaknesses of a product, carefully worded probes are one of the keys to both uncovering and gaining insight into issues.

Probes occur for many reasons, such as when you want to

- be clear about what participants are thinking.
- find out if participants understand a concept or term.
- understand why participants chose one option or one path over another.
- know if an action or outcome was expected or not.
- ask participants about nonverbal actions (e.g., a squint) or sounds (e.g., a sigh).
- find out if participants saw an option, button, link, etc.

6.3.1 **Probing questions**

Keep in mind that asking questions often puts people in a defensive position. There is a technique to avoid making participants feel challenged. There are two parts to the technique: the linguistic part and the tone-of-voice part. First, instead of asking a question, use a "curious command," which is linguistically an imperative but sounds neither like a question nor a command (Mitchell, 2007). These statements start with words like "tell me" and "explain" and need to be said with an empathic tone of voice. Here are some examples of phases you can use.

- Tell me a little more about . . .
- Describe a bit more about . . .
- Share some more about . . .
- Talk some more about . . .
- Help me to understand a little about . . .

6.3.2 **Planned versus spontaneous probes**

There are two types of probe: planned and spontaneous. When you know ahead of time that you will want to get additional information at a certain point in a task, the probe is planned. Usually the decision to probe is made during test planning. For example, let's say a web site design team is trying to decide whether users, after entering information into a dialog, will want to stay in the dialog (which means using an Apply button) or leave the dialog and go back to where they came from (which requires an OK or Done button). The design team decided to use OK, but they want to know if this works for users. So after participants click OK, the moderator asks, "Is that what you expected, or not?" A discussion of the options can then follow.

Sometimes the decision to ask a planned probe is not made until the pilot test or even after one or more sessions have been run. Whenever the requirement for a probe appears, write it down in a place that will help you remember when and what to say. You can add the reminder to the test script, for instance, or put it on a separate piece of paper kept with the other materials you're using (scripts, task cards, checklists, etc.)

Unplanned probes can occur at any time during a diagnostic test. Whenever you think you need more information or that the participant has more to add, a probe may be appropriate.

6.3.3 **Probes to avoid**

Always keep the importance of avoiding bias in mind when asking questions. In addition to the rules discussed in chapter 4, the following two guidelines about probing are critical.

Don't talk over participants or interrupt them with a probe. You don't want them to forget what they were saying or doing. If you both start speaking at the same time, defer to the participant.

Avoid using a probe to talk indirectly to developers through participants. For example, if a participant says, "This screen is too busy" and you respond, "What did you say?" even though you heard the comment, you may be trying to talk indirectly to developers through the participant. You are trying to get the participant to make the statement louder or to elaborate on the negative comment. A subtly different probe is, "Tell me more about what you mean."

6.3.4 **Common probes**

Here is a list of probes we have found useful.

- Tell me what you think about that task.
- Is that what you expected, or not what you expected?
- Did you notice the [name of a UI object], or not notice it?
- What would you do next?
- You just said, ["participant quote"]. Help me to understand what you meant by that.
- I noticed that you paused before clicking [name of a UI object]. Share with me what you were thinking at that point.

6.4 **PROVIDING ENCOURAGEMENT**

When poorly designed products are tested, participants make errors, need assistance, and fail some tasks. Since you likely helped plan the test and conducted a pilot session, you can probably predict fairly accurately when participants will struggle with the product. Unfortunately, most participants blame themselves when they struggle, even when you have told them not to.

Providing encouragement is one way to try to ease the burden on participants. Encouragement helps keep them motivated to continue. The way you provide encouragement is important, however, because you don't want to be in conflict with your role as a neutral observer. A good practice to follow with every participant is to pick a task early in the session, say task 3, and always say something encouraging *after* that task. For example, you might say, "Your thinking aloud is very clear. You're being very helpful." Remember to say this in an engaging rather than an automatic tone.

Why *after* the task? As a general rule, you don't want to reinforce either positive or negative behaviors. The safest way to give encouragement is to separate it from task performance or a specific statement or action during a task because some participants may take your encouragement to mean that you want them to make more of those kinds of statements or actions.

6.4.1 **Encouraging statements to avoid**

Imagine a situation where a participant has just failed three tasks in a row and then completed the next one. It's very tempting to intervene to make the participant feel better by saying something like "That was great!" or "Good work on that one!" The problem with this kind of encouragement is that some participants may misinterpret it as meaning that you want them to succeed because you want the product design to succeed. Try using one of the more neutral statements in the next section.

Of course, as a general rule you don't want to provide encouragement during a task; however, there are times when it's called for, such as when participants are getting discouraged and frustrated. It does no good to maintain your neutrality if participants lose their motivation to continue.

6.4.2 **Common encouraging statements**

Here are some encouraging statements you can use both within and between tasks.

- You're doing fine.
- You're really helping us.
- You're giving us the kinds of information we need to make this product better.
- Your thinking aloud is very clear and helpful. Thanks.
- Don't forget you're helping future users by working with me today.

6.5 **DEALING WITH FAILURE**

By its very nature, usability testing is meant to uncover design flaws—places where the product interface doesn't match users' abilities, experiences, or expectations. Therefore, you *want* users to encounter problems so that the problems can be fixed. Unfortunately, users often feel that they have failed if they can't figure out how to complete a task. This is one of the most difficult parts of moderating test sessions.

■ **FIGURE 6.1** A participant's body language shows signs of distress.

Our Typical Instructions

We use the following statements when participants seem disturbed because they didn't complete a task.

"Remember that we are not testing you or your abilities in any way. We are just trying to find out where the product causes people confusion so that we can make it better. If you have any issues with the product, that actually helps us because it shows that the product wasn't clear to you. We can then fix that problem so that all the people in the future won't have to deal with it.

"We are going to ask you to complete a number of tasks with the product. Keep in mind that not everybody completes all the tasks; that's OK. Sometimes I may have you move on so we can get to other tasks. Just do the best you can."

6.5.1 Participants' self-blame

Unfortunately, even if participants understand and believe that you are not testing *them*, feelings of failure usually begin the first time they have trouble, and that's when they might make statements to defend themselves.

We've noticed that this is particularly true when participants are less experienced with the product being tested, be it a software application, web site, medical device, or consumer product. Novice users will assume it's their lack of knowledge or expertise that's causing them to fail, whereas experienced users more often recognize a poorly designed interface when they see one and are less likely to blame themselves.

Typical self-blame statements include:

- Maybe it's just me.
- I can't believe I didn't see that!
- Well, normally I would have read the entire user manual.
- That was my fault. I didn't read the sentence at the top of the screen.
- If I had more experience with this product, I probably would have seen that button.

This is important: You shouldn't make the product look better than it is by taking failure away or by minimizing its impact, such as by saying things like, "Everyone has trouble with that task." We show this scenario in Video 3 on the book's web site.

6.5.2 The moderator's distress

Remember that after the test begins, the moderator's responsibility is to be a neutral observer. That role can be a challenge. Most of us would not let someone have an unpleasant experience while we watch and remain silent, and yet that's sometimes what we have to do. Giving participants a very difficult task may make you feel that you are setting them up for failure. As a moderator, you must suppress your strong desire to take away or minimize the hurt you believe that participants are experiencing.

This is when novice moderators are most likely to intervene inappropriately and say something that biases the participant. Their intuition screams, "Help them!" Ironically, those who become usability professionals are usually very people oriented; they're tuned into other people's feelings. Outside of the testing situation, they would take action to help others in distress. Becoming an effective moderator means learning that sometimes you need to suppress those feelings or, at least, not to act on them.

Those new to moderating also have a tendency to project their feelings onto participants without realizing it. A three-minute struggle by a participant can seem much longer to someone who is watching!

The anxiety and/or frustration that observers feel may or may not be of the same intensity as the participants' feelings. For example, when we've demonstrated how to moderate test sessions to graduate students, we've noticed that the student observers actually get more emotional than the participants. The participants have negative feelings when they struggle, but they're also busy trying to complete the task. They are *working* as well as *feeling*, while the observers have little to do but feel. When participants are asked after these test sessions about their feelings, their expressions of emotion are usually much more muted than those of the observers. With practice and time, you will be able to avoid projecting your feelings onto participants and remain calm.

6.5.3 **The participant's distress**

The point at which participants become so upset that they can't continue to be effective will vary. Also, the interpretation of when that point is reached varies across moderators. We've been unable to find a clear, simple rule for determining the level of participants' stress. We're also reluctant to overrule the intuition of moderators when it differs from our own. We can only tell you the cues we look for and what we do when we see them.

The cues that tell us that participants' emotions may have become too intense are

- sweating, especially around the face and neck
- looking intensely at the product and nowhere else
- tension in the voice, or sighs, or huffs
- obvious anger
- silence for long periods of time
- tunnel vision—an inflexibility in exploring options and repeating the same things over and over
- tears or rapid blinking

Notice that most of these cues are nonverbal expressions of distress. Some of these signs are shown in Video 4 on the book's web site.

You can sometimes tell when participants are just trying to convince you to stop the task but are not in distress. They may give verbal expressions of distress but show none of the listed signs. In those cases, we usually ask them to continue.

If a participant expresses verbal distress along with additional cues, your first response should be to tell them they're providing valuable information and to remind them that they are helping future users by what they are finding. But it may not help because they clearly don't see it that way. They usually expect to perform much "better" so they see themselves as failing. What can you do in these cases?

6.5.4 **Responses to participants' distress**

When you see signs of distress, you need to take some action to either assess the participants' emotional state or reduce the intensity of their feelings. Here are some steps you can take if you suspect the tension is too high.

When you're in the room with the participant:

- Say, "How are you doing?" This helps you assess his or her state.
- Use the participant's name to provide extra emphasis on what you say.
- Take a break. In this situation, it's often better if you say *you* need a break because if you ask the participant, he or she might say, "I don't need one."
- Have the participant stand up, get out of the seat, and walk around if possible. Ideally, take the person out of the room to get water or coffee. It's important to change body position and get some air.

During the break, get some refreshments and talk about other things. Watch how the participant is reacting. Use your intuition to sense whether he or she is tense or preoccupied. Tell participant that the feedback being provided is very valuable.

If you sense that the stress level is too high, it's best to ask directly about it. Say something like, "Please tell me if you are finding this session too difficult" or "Help me understand how you're feeling about the session." At this point, it doesn't do any good to try to remain neutral in order to preserve the integrity of the data. If the participant is very upset, the data might not be valuable anyway.

When you're moderating from outside the room:

- Take a break and go into the room and sit with the participant or bring the person out of the room.
- If necessary, volunteer to sit with participant as he or she works. This may help some feel calmer.

When you are conducting a remote session, keep in mind that distress is often accompanied by silence. Many of the nonverbal signs of distress depend on visual contact. When there is silence in a remote session when the participant is struggling, you need to find out whether he or she is just forgetting to talk or whether the person's really upset. Try asking, "Are you finding this session difficult?" This is a way to open up the topic for discussion.

If the participant is stressed but you sense that he or she can continue, pay attention and look for signs of escalation. If you have the freedom to do so, consider changing the script a little, perhaps skipping the most difficult tasks, to avoid increasing the participant's stress level.

6.5.5 **Stopping a test**

If your intuition tells you it's the right thing to do, ask the participant if he or she wants to stop the session. Sometimes when we stop a session early, we conduct the post-test interview to wrap up the session and reinforce to the participant that he or she provided valuable feedback, even though all the tasks weren't completed. In other cases, if the participant's stress starts to escalate again, we simply say, "That's great. Why don't we stop here. I have a few questions for you about your overall experience with the product." This allows the participant to "save face," and he or she may not even be aware that the session ended early.

Note: If there is an incentive and you ask the participants if they want to stop the session, don't initially bring up the fact that they will be compensated even if they stop. This sometimes makes participants feel worse about themselves and they may say, "I think I will stop, but I am not taking the money." It is better to establish if they want to stop first, then later simply give them their compensation with a "thank you." If they protest, you should take the responsibility off participants' shoulders by saying, "We compensate everyone who takes their valuable time being in a session."

6.5.6 **What you shouldn't do when a participant fails**

Commiserate. When participants fail, your first reaction may be to make them feel better. It's very tempting to say, "I can see why that was hard." It's also tempting to nod your head in vigorous agreement when participants say things like, "Wow, that took forever," or "That was really confusing." Our advice is not to commiserate. Acknowledge that you heard and understood the comment, perhaps with a little nod, and say, "OK" or "Mm hmm."

Tell a white lie. You may also be tempted to reduce the impact of failure by telling a "white lie," such as, "Don't worry, everybody has trouble with that task." (This occurs in Video 3.) In some cases, the statement is even true. The problem is that you're reinforcing participants' feelings, which may increase their level of negativity toward the product and possibly affect their future performance, ratings, or comments.

INTERVIEW WITH AN EXPERIENCED MODERATOR

What do you do when a test participant has a strong negative reaction to the product you are testing and is ready to stop the session?

During a usability test I frequently sense that the participant is having more difficulty than they are verbalizing. It is always a challenge to sort out what a participant is doing from what they say they are doing. In fact, I have occasionally had the secret desire to stop a test and tell the participant that I will pay them double if they tell me what they are really thinking!

I did encounter a person who wanted to tell me what she really thought and it took me by surprise and caused a few uncomfortable moments for both of us. Here is what happened.

The participant started performing the tasks as usual. However, she became more and more frustrated. Finally she just exploded with annoyance, telling me that the product was not any good, that she was frustrated, and that she did not think she wanted to continue.

My first reaction was to become defensive. I wanted the test session to be pleasant and congenial. I realized, at that moment, that I wanted her to enjoy herself!

But she was doing just what I always wanted a test participant to do—displaying her honest reaction to the product. I had to respond quickly to see if I could make a productive session out of a potential disaster. As soon as I realized that I had a frustrated but honest user, all I had to do was to turn her from annoyed to constructive.

We stopped the tasks ("Let's stop for a minute and talk about how this is going.") I acknowledged her frustration ("I heard you when you said you were frustrated.") We talked about how she was giving me just the feedback I needed to improve the product and that her experience was important to me. ("It is really exciting to see you react in such an honest way. If you can take a minute and describe to me what you are thinking and what is challenging about the task, I will understand more about what could be improved in the product.") With her emotions validated and permission to talk honestly, she quickly became more specific and was able to tell me what she did and did not like. I kept her talking and giving honest feedback and encouraged her that her honest feedback was more valuable to me than platitudes.

Lessons learned? (1) Be careful what you wish for. (2) When what you wish for comes along, be ready for it. (3) Be prepared to turn frustration into constructive feedback.

6.6 **PROVIDING ASSISTANCE**

Usability problems often reveal themselves sequentially as participants move through the steps in a task. A problem that occurs early in the sequence and stops progress can prevent the participant from uncovering additional problems that would come later. Providing assistance is a way to move past one step so that later problems might be uncovered.

Unfortunately, when you give participants assistance, you may provide them with information that they wouldn't have if you didn't help. Your intervention could change how participants perform and also perceive the product. Because of this, you should give more assistance in diagnostic tests (where finding usability issues is the primary goal) and less in benchmark and comparative tests.

Providing assistance is one of the most important skills of moderating. It requires balancing two factors: helping participants along while avoiding giving information that will help them complete later tasks.

6.6.1 **An assist**

Let's look at a specific example to clarify what an *assist* is. (Video 4 on the web site provides an example.) Imagine that participants in a diagnostic usability test are asked to complete a task that requires them to locate a dialog box by selecting a menu option and then setting some controls in the box, such as unchecking a box or selecting a radio button. The product team suspects that there are several usability problems with the design that may make this task challenging: It may be difficult to (1) find the correct menu option to open the dialog and (2) set the controls in the dialog box once it's open.

The moderator would like to gather data on both locating and using the dialog. Consequently, when the participant cannot locate the correct menu option, the moderator tells him or her where it is and then lets the participant set the controls in the dialog box (the moderator then records the assist).

6.6.2 **Giving assistance**

One clue that may call for you to provide an assist is seeing a repeating pattern of unproductive behavior. But some other situations may call for an assist, such as the following:

- The participant has tried several alternatives and asks for help.
- The participant is heading toward a sequence that will cause a product failure or system crash.
- The participant is approaching the task time limit or is taking so long that he or she won't have time for later tasks.
- The participant thinks the task is complete when it's not.

Perhaps the most difficult situation is providing assistance that *is* likely to help participants complete subsequent tasks in addition to the current one. This tends to happen when participants encounter global issues, ones that occur across the whole interface rather than as isolated instances. Some times you have no choice but to give the assist so the participant can continue. You can handle this problem later in your data analysis.

One factor in deciding to give an assist is your relationship with developers. If the developers are likely to agree that there is a problem and they will fix the problem, then it's best to give assistance in order to uncover more issues. On the other hand, if you feel that the developers need to see repeated failures to motivate them to make a change, then it's best not to give assistance. The consequence of giving no assistance is that you will miss some subsequent problems. This situation is one of the trade-offs of moderating in the real world!

One last point to consider: A complex product that provides tools for data manipulation or analysis by a skilled analyst may require some trial and error to learn. Consequently, providing assistance may not indicate that there is a major design flaw. In addition, complex analytical products are typically used by people who are good at problem solving and who are not easily put off by a challenge. In this case, your giving an assist may not change their perception of a product's usability in the same way as it might in a self-service product, so you can be freer in stepping in.

6.6.3 **Interventions versus assists**

Some interventions are not assists. For example, prompting participants to think out loud (described in section 6.3) and providing encouragement (described in section 6.4) are not assists.

It's not an assist when you clarify a task. For several reasons, participants sometimes begin performing a different task than what you intended. It's standard practice in those cases for moderators to immediately intervene to make sure that participants understand the task. A common strategy is to have participants reread the task scenario and, perhaps, say it in their own words. If participants still don't understand the task, you may restate it in different words until they do understand, which is not an assist.

It's not an assist when you help participants recover from a bug or redirect them when they choose a correct but unanticipated path. Sometimes participants do understand the task, but they choose a path to completion that you didn't intend. For example, a carefully prepared scenario to force participants to move three levels down in a menu structure is suddenly negated when a participant finds a right-click, pop-up menu that performs the same task. This could be a surprise to you if you didn't know there was a second path to success. You can intervene immediately or wait for the end of the task and then say, "That was a correct way to do that task, but there is another way. I'd like you to find it and then do the task again." While this is an intervention that may change the participant's perception of the product, it is not considered an assist.

6.6.4 **Not giving assistance**

Whenever participants are moving toward a solution to the task, even if they're going down the wrong path initially, there is no need to give an assist. Unless they are clearly stuck, resist the temptation to jump in and help!

For some usability tests, the decision is made during planning not to provide *any* assistance for the entire test or for a particular task. For example, in a competitive test of two or more products, it is almost impossible to provide assistance for one product that will not favor that product over its competitors.

A common occurrence, especially when participants encounter the first difficult task, is for them to ask for assistance or say they don't know what else to do. The decision to assist depends on whether you believe that participants have worked long enough to have exhausted their options and/or are repeating similar patterns. If you decide the answer to both questions is no, then instead of an assist, you might say, "Let's keep working a bit longer" or "Why don't you see if there's something you haven't tried?" One reason for asking participants to keep working is to avoid giving them the impression that they can end any task they find difficult by saying they don't know what else to do.

An additional factor that can influence whether to give an assist is the type of product you are testing. Some products are intended to be self-service—people are supposed to be able to use them without a manual or training. You should not need assistance, for example, to look up your company's benefits package on its web site or take money out of an ATM. Consequently, if participants need assistance to complete tasks with a self-service product, it usually indicates a design flaw that must be fixed.

6.6.5 **Levels of assistance**

The levels of assistance move from general to specific, and from providing as little information as possible to telling the participant exactly how to complete a task. Keep in mind that as you move through this sequence, you can decide at any point to terminate the task and move on to the next one, or you can skip earlier levels and go to the later ones.

Level 1. Breaking a repeating sequence. When a participant continually repeats the same or similar sequence several times, all it takes sometimes is a change in concentration to get them unstuck. A simple "So, what do you think is going on here?" or "Try reading the task again" is often enough to return their focus to the goal. Some practitioners might not call this an assist, but we believe it is because it is an intervention that would not occur if the moderator were not present.

Level 2. Providing a general hint. Often participants come close to finding the option they need. For example, they may have opened the correct menu but not read far enough down the list of options or scanned them too quickly and missed the correct one. You could provide a level 2 assist by saying, "Remember how you started this task? You were getting close." or "You actually went by the option you need." If they did not get close yet, you might say, "Sit back and take a look at the whole screen" or "It's in a menu you haven't opened yet."

Level 3. Providing a specific hint. When a level 2 assist does not move participants along, you may have to be more specific. For example, you could say, "The option is in the Edit menu" or "Try all of the options in the list." An assist at this level focuses participants' attention on the relevant area but still allows the moderator to obtain information about things like the terminology of options. And it still requires work on the participant's part.

Level 4. Telling participants how to do the next step. In some situations, you may decide to tell participants how to perform the next step. For example, "Open the Edit menu and select Preferences" or "Click the third bullet." This type of "next step only" assist standardizes what an assist is. This standardization can become important when it comes to counting assists. It's also important if there are several moderators because you want to make sure that all are consistent in how they give assists.

6.6.6 **Completing a task for a participant**

Sometimes a task must be completed to allow participants to attempt later tasks. These are often referred to as "contingent" tasks. Usually you can avoid contingent tasks with careful test planning, but sometimes they're unavoidable. This situation can be awkward for both you and the participant because it can make the participant feel like a failure. Also, if you don't handle it properly, you could provide participants with information that they wouldn't have on their own.

If you would reveal important information by completing the task, you can ask the participant to look away while you complete it. For example, you might say, "Could you look out that window while I do something on the computer?" If it's going to be obvious that you're completing the task, it's best to be straightforward and say something like "I need to complete this task so we can go on to the next task. So if you could look away, I am going to finish it."

New moderators tend to have difficulty with this situation. They are tempted to try to make the situation less awkward by elaborating: "Don't worry, everyone has trouble with this task." As we discussed earlier, such statements can bias the test by influencing the participant's perception of the usability of the product. It would be better to just say, "Remember, don't blame yourself when there isn't time to finish a task."

6.6.7 **Measuring assists**

It's beyond the scope of this book to discuss the measurement of usability. But we do want you to recognize that counting assists can be complicated. You can see from our discussion that all assists are not the same and that the decision about providing assistance depends on a number of factors. So it's not easy to decide whether providing an assist equals task failure or whether it takes several assists to count as task failure or whether a task completed with several assists should count as a success.

Some organizations compute and report task success statistics on their usability tests. These statistics allow them to see trends in the quality of their designs over time and determine which parts of the organization are producing more usable products. For example, they might compute the percentage of tasks completed without assistance and the percentage completed with assistance. To make those values meaningful, it must be clear what an assist is. One way to be consistent is to restrict assists in summative tests to the specific "next step only" level 4 assist. Consequently, an assist always means that the moderator told the participant how to perform the next step in the task. This strategy makes the measurement clearer.

Our point is that before the test it's important to carefully consider when and how you will provide assistance and how you will count assists afterwards. If several people will moderate test sessions, it's even more important to plan ahead so that assists will be handled consistently.

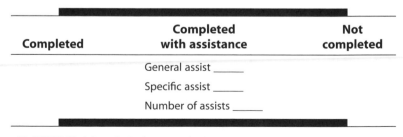

Completed	Completed with assistance	Not completed
	General assist _____	
	Specific assist _____	
	Number of assists _____	

■ **FIGURE 6.2** A data collection form to record assists.

Figure 6.2 is a sample of a data collection form for recording task completion rates as well as the types of assistance provided. The moderator circles one of the three outcomes (Completed, Completed with Assistance, or Not Completed) and then if assistance was provided, he or she checks off the type(s) of assists.

WHAT THE RESEARCH SAYS

Do some usability problems hide other problems?

Usability problems often reveal themselves sequentially as users work through a task. An interesting study provides some data to support this (Medlock, Wixon, Terrano, Romero, & Fulton, 2002). In this study, the authors used what they called the Rapid, Interactive Test and Evaluation (RITE) method. A RITE test is a variation on a traditional usability test. It is planned the same way. The difference is in the way it is executed. A team of people including a usability specialist, product developers, and key product stakeholders attends and watches the first session or sessions. After watching them, the usability specialist lists the usability issues that occurred and the team discusses them. If the team is in agreement about the problems and appropriate solutions, then the problems are fixed immediately and then the next session is run. If they want to be more confident about the issues, they may run more test sessions before making changes. After making changes, the team verifies that the solutions are working and notes whether there are any additional problems.

Notice that with this method the focus is on finding solutions to problems and verifying them rather than just finding the problems. In the study we are discussing, the number of failures and errors actually *increased* after the first set of solutions was applied. You might think that the cause of the increase was poor fixes to the problems. But the increase actually was caused by the fact that the first session revealed a few major problems that kept participants from making further progress on tasks. When the small number of big problems was fixed, nearly twice as many errors and failures occurred because the participants were able to explore much more of the user interface. Consequently, the team kept the first set of fixes and applied additional ones to the new problems. After running about 10 sessions almost all of the errors and failures had been eliminated.

Medlock, M. C., Wixon, D., Terrano, M., Romero, R., & Fulton, B. (2002). Using the RITE method to improve products; a definition and a case study. Paper presented at meeting of the Usability Professionals' Association, Orlando, FL.

Chapter 7

Interacting during post-test activities

The time between the final task and the end of the session is very valuable. Task performance may be over, but you still need to collect important measures of participants' subjective perceptions. Participants have spent perhaps an hour or two with the product, so it's important to capture their perspectives and to take the time to clarify what happened during the session.

In a diagnostic test, this time provides a last chance to uncover product strengths and weaknesses. In a summative or comparison test, it's a time to gather measures that summarize the users' experience.

In this chapter, we discuss the activities that happen after the test itself:

- maintaining your roles
- determining the order of activities
- clarifying things that occurred during the test
- administering ratings and questionnaires
- asking open-ended questions
- allowing others to interact with participants
- providing incentives and other final activities

7.1 MAINTAINING YOUR ROLES

When the last task is over, there's a feeling of relief for both you and the participants. This is understandable because working on tasks is almost always the most stressful part of the session. Usually the more stress and failure participants have experienced, the more relief both parties feel.

Although it's appropriate to ask participants to relax for a few minutes, there is still important information to be gathered. As a moderator, you need to maintain your roles until the participant leaves. You might say, "Let's relax for a minute before we talk about your reactions to the product." This statement makes it clear that the relaxation and the talking are separate.

As the Gracious Host, you can ask participants if they want to take a break. If you sense that participants really need a break, don't leave the decision in their hands. Tell them that *you* need a break. If you're sitting with them, get up and offer the participants whatever refreshments you have. You might also stretch and encourage them to do so, too. If you're in another room, say, "Relax for a moment. I'll be right in." In a remote session, say, "I need a minute to stand up and stretch. You may want to do the same."

As the Neutral Observer, you don't want to bias participants' perceptions by talking about how hard or easy the tasks were to do. Making leading statements should *not* be part of relaxing. Likewise, statements about relaxing should not be tied to specific tasks or the success or failure of participants. The break is to relieve fatigue from concentrating for a fairly long time.

In your role as Leader, the break is often a good time to talk with any visitors about whether they have additional questions they want you to ask participants. If the questions they want to ask are leading or use jargon, change the wording. For example, the visitor might say, "Ask her if she liked the Customization option." You might then ask the participant, "What do you think about the feature that allowed you to reorganize the sections of the screen?"

7.2 DETERMINING THE ORDER OF ACTIVITIES

The activities in the post-task session vary depending on the goals of the test. There are at least four common possibilities:

1. A discussion about key events that happened while participants were working on tasks.
2. A set of open-ended questions about the product.
3. A set of questions about aspects of the product that were *not* covered by the test (e.g., questions about their previous use of the product or about their experiences with a competing product). As the relationship between market research and usability has increasingly overlapped, we find that our clients often want to ask

marketing-related questions after participants have used the product, such as how likely they would be to purchase the product.

4. A self-administered questionnaire or a set of closed-ended questions or ratings that ask about the usability of the product. The questionnaire might be one of the standardized options that have been developed—for example, the Software Usability Scale (Brooke, 1996)—or one that your organization has developed internally.

When deciding the best order for post-task activities, the general rule we follow is:

If the test is diagnostic, first perform the activities that will tell you the most about the strengths and weaknesses of the product. If the test is summative, first have participants perform the activities that measure perceptions of usability.

In a diagnostic test, first discuss key events that happened while participants were attempting tasks, then move to subjective ratings and open-ended questions. You want to uncover usability issues that come from the participants' reactions to the product and their performance.

In a summative test, first administer questionnaires and ratings. Typically, these measures are more reliable the closer they are to the events they refer to. In a comparison test in which participants have used more than one product, you may want to know which one they prefer. We suggest you ask that first, so that discussion of open-ended questions and task performance doesn't change their ratings or answers.

Of course, there are tests that have more than one purpose. For example, a test may occur near the end of the development cycle with a measurement focus, but it also may be the only test performed on the product. Consequently, the test team may be looking for any diagnostic information it can get in addition to measuring usability. In those cases, you need to prioritize the objectives with the team and jointly decide on the order of post-task activities.

7.3 CLARIFYING THINGS THAT OCCURRED DURING THE TEST

In a diagnostic test, the goal is to extract information that will help you and the team uncover usability issues. Often, events have occurred at times when you didn't want to interrupt the participant for clarification. Now is the time to ask about those events. Some issues are best discussed after participants have seen all of the product or worked through all of the tasks. For example, in tests conducted early in the development cycle, you often want to know if participants understand the underlying concepts of a user interface. Do they "get" the concept, part of it, or none of it? These issues usually are more important than quantitative subjective measures.

In a purely summative test, the focus is usually on measurement, and clarifying issues may be less important. In tests that are aimed at *both* diagnosis and measurement, it is important to collect the subjective measures first and ask clarifying questions second.

7.4 ADMINISTERING RATINGS AND QUESTIONNAIRES

It is common to ask a set of closed-ended questions and a set of ratings at the end of a usability test. The closed-ended questions may be unique to the product tested or may be from one of the standardized questionnaires published in the literature. Some organizations have a set of ratings that they use in every test to provide a basis for comparison over time.

7.4.1 Questionnaires

One of the decisions to be made during test planning is whether to have the questions self-administered by the participant or read by the moderator. There is an interesting trade-off involved in this decision. When the questions are self-administered and responses are written, they are not part of the recording and any visitors or people watching a recording can only see the answers after the session is over (unless you have an overhead camera and can zoom in on the rating sheet, of course). There may be valuable information in the answers that developers only see when the report appears. On the other hand, participants may be more candid when they answer the questions in writing.

There is, however, a third possibility, which is to have participants fill out the questionnaire on their own but then go over the answers out loud with the moderator. Video 5 shows this approach. Some moderators who viewed Video 5 reacted negatively to this practice because they see it as pulling an unexpected surprise on participants—participants thought their answers were private, but then the moderator discussed them in public. If you're concerned about that issue, you can tell the participants exactly what will happen. For example, "I would like you to fill out this questionnaire. Then I'll come back so we can discuss your answers."

When the questions and ratings are self-administered, it is best to leave the test environment or to be silent or take a break in the case of a remote test. You don't want the participant to feel rushed. In addition, some moderators feel that participants are more candid when they are alone.

7.4.2 **Reasons behind the ratings**

In a diagnostic test, the value of the answers to the questions is in the light they shed on the strengths and weaknesses of the product. A typical diagnostic test with five to eight participants cannot provide reliable quantitative data on subjective measures. Consequently, the important information is *why* participants chose their answers rather than the numeric values they assigned.

For example, the important thing about a rating of 6 out of 7 is not the number itself but why participants chose that value. Therefore, you always need to ask "why" when the participant doesn't provide that information. You might ask, "So what does a 6 mean here?" As long as you ask this question for all of their answers, participants should interpret your question as clarifying their opinions rather than putting them on the spot.

WHAT THE RESEARCH SAYS

Does it matter which questionnaire you use?

Tullis and Stetson (2004) conducted an interesting study comparing questionnaires used at the end of a test session. They tested five questionnaires including the Software Usability Scale (SUS) (Brooke, 1996) and two other standardized questionnaires. The Software Usability Scale has 10 rating scale items and was developed by a group of usability specialists at Digital Equipment Corporation. Tullis and Stetson also included a questionnaire they developed themselves and had used in all of their tests.

They asked 123 participants to perform two tasks, each on a web site. The participants were then given one of the questionnaires with which to evaluate the two sites. The web sites turned out to be quite different in their perceived usability, with one site being statistically more usable than the other over all of the 123 participants.

The authors then took samples of gradually increasing sizes to see how each questionnaire discriminated between the sites. For example, with a sample size of 6 participants, could a questionnaire show that one site was more usable? How many of the samples showed this difference? They then took many samples of size 8, 10, 12, and 14 and computed statistics to see how many of the samples showed the difference in usability between the sites.

The results showed that *none* of the questionnaires showed a difference between the sites at better than an 80 percent rate with sample sizes of either 6 or 8 participants. With a sample of 10 participants, only the SUS had 80 percent of its samples showing a difference, and only the SUS achieved a 100 percent rate with sample sizes of 12 and 14.

The bottom line is that the SUS was more sensitive with small samples. Their locally developed questionnaire was just average among the rest of the questionnaires.

In addition, this study shows that none of the questionnaires discriminated well with sample sizes of fewer than 10 participants, which means that questionnaires have low reliability with the sample sizes we typically use for usability tests.

Tullis, T., & Stetson, J. (2004). A comparison of questionnaires for assessing website usability. Paper presented at the Annual Meeting of the Usability Professionals' Association, 1–12.

Often participants give a rating that's between two points, such as, "I would give it a three and a half." You can then ask, "Why between those values?" then later, "If you had to pick one number, which would it be?"

7.4.3 **Accuracy of ratings**

One of the main issues about ratings is the truthfulness and reliability of the values participants give. Frequently, participants give higher— that is, more favorable—values than the moderator and team think they should. It's common to watch participants struggle with tasks, then at the end of the session give the product a favorable rating of 6 out of 7! The testing literature talks about this issue and attributes it to the demand characteristics of the situation, that is, factors other than participants' task performance that could push up ratings. These factors include the need to please, the need to be positive and not critical, and the need to appear computer or product literate. Sometimes participants have an easy time with the last task and assume that they were "getting it." In other cases, participants may really want and need the product or feature you are showing them. Even though it may not be usable, the *functionality* is very beneficial to them. As a result they don't want to say anything negative because they fear it may delay the release of the product or feature.

There is no research verifying or clarifying the exaggeration of ratings at the end of a test. We have run demonstrations of testing sessions with graduate students and seen this effect. When we ask student participants why they gave the values they did, they say something like, "Well, when I thought about all of the products I have used over the years, this one was not that bad." They're saying that they have a good reason for their ratings and that they were not conscious of pushing up their ratings.

Our impression is that ratings given after each task are more valid than ratings given at the end of the session. Post-task ratings seem to be consistent with other measures; that is, when participants struggle, they give the task a lower rating.

There are ways to help participants understand what it is you want them to rate. Instead of just saying, "What is your overall rating of the product?" you can be more specific. For example, you can say, "I want you to rate your experience with the product over the past X minutes on the scale I am showing you. Don't rate it on the basis of how it might be to use in the future or how it might or might not work for other people. Just rate it on the basis of your actual experience today."

7.5 **ASKING OPEN-ENDED QUESTIONS**

Open-ended questions often are asked at the end of a post-test interview. The standard closing question is, "Is there anything more you have to add?" or simply, "Anything else?" Our experience is that participants usually interpret those questions as an invitation to be done with the session. So save them for what you think of as the very end of a post-test interview.

There are other questions that provide participants with an opportunity to summarize their experiences. Common ones are, "What three things did you like best and least about your experience with this product?" and "You have been working with this product for about an hour. Tell me about your impressions about its ease or difficulty of use." Those types of question often elicit valuable comments that you can quote in a report of findings or in video highlight clips. In addition, if you ask the same question of all the participants, you can make statements in your reporting such as "Five of the six participants listed the autofill feature as a valuable addition to the data entry screens."

One common problem is that participants only comment on the functionality of the product and not its usability. For example, when asked for the three things they liked best, participants might say, "I liked that you can create reports." That answer may be valuable, but it comments on the presence of a feature, not its ease of use. Video 5 shows an example of this confusion. You might follow up with, "OK, but was it easy or hard to create them?" You should make it clear in your initial question that you are asking about usability.

Figure 7.1 is an example of a self-administered, post-test questionnaire created by a test team. Figure 7.2 is an example of a list of post-test interview questions asked verbally by the moderator.

7.6 **ALLOWING OTHERS TO INTERACT WITH PARTICIPANTS**

When the post-test interview is over, there is no more data to collect. Some moderators then ask developers or managers if they want to speak with participants. The invitation can occur in both local and remote sessions, and visitors often take that opportunity. Sometimes they talk to participants about something that they saw during the session; other times they want to show the participants new features. Our experience is that both visitors and participants enjoy the experience. Many developers never get to talk to end users because they're not allowed to go on customer visits.

That said, it's important that you monitor the conversation, especially when you're unfamiliar with the visitors. Make sure the interaction stays on a professional level. You are still responsible for the ethical treatment of participants. In our experience, almost all of these conversations are friendly and informative on both sides. It's rare, but sometimes visitors ask personal questions or ask participants to work on more tasks or make promises about the product that they have

Self-Administered Questionnaire

How was your overall experience with the product, on a scale from
1 *(worst)* to 7 *(best)*?

     1   2   3   4   5   6   7

Why?

Did your impressions of the product change after using it?
    ☐ Yes   ☐ No

If so, how and why?

How would you describe the differences and/or similarities between this
product and other products you have used?

Please circle the two words that **best** describe the product:

Thorough	Confusing	Valuable	Difficult
Amateur	Unimpressive	Shoddy	Complete
Overwhelming	Efficient	Easy to use	Ineffective
Helpful	Professional	Deficient	Time-saving
Intuitive	Problematic	Sloppy	Expert
Fool-proof	Quirky	Rewarding	Unreliable
Reliable	Questionable	Trustworthy	High quality

■ **FIGURE 7.1** A post-test self-administered questionnaire.

Post-Session Interview Questionnaire

1. How does this web site compare to other similar ones you have seen?
2. If you could make one significant change to the website, what would it be and why?
3. What do you like **most** about this web site?
4. What do you like **least** about this web site?
5. Is there anything you would add to this web site if you could?
6. Is there anything you would remove from this web site if you could?
7. Would you recommend this web site to a friend?
8. Do you want to add anything else?

■ **FIGURE 7.2** Post-session interview questions a moderator might ask.

no control over. If you hear anything that bothers you, just say they need to stop so you can get ready for the next session or clean up the lab.

7.7 **FINAL ACTIVITIES**

There are a few more steps before you end your interaction with participants. Many tests provide an incentive or other compensation. You not only need to provide it but also must document that you provided it by having participants sign a receipt. The documentation may or may not be required by your organization, but it's important to have if any participant later claims he or she didn't receive the incentive. It's rare, but it does happen.

■ **FIGURE 7.3** A participant receives her incentive.

7.7.1 **Providing incentives**

These four rules apply to giving test participants the promised incentives after local tests.

1. If the incentive is cash, provide it to participants in an envelope, have them count it, and then have them sign a receipt saying they received the cash and the amount.

2. If the incentive is a gift certificate or gift card, provide it in an envelope and tell participants what they have to do to use or activate it. Again, have them sign a form saying they received the item and record the number on the card.

3. If the incentive is a check or merchandise, such as a product your company makes or a garment with a logo, have participants sign for it.

4. If the incentive will be sent to participants later, have them complete a form with their contact information. Read the form to make sure it's complete and that you can read it. Also give them a copy of the form or a different sheet indicating when they should expect payment and whom to contact if they don't receive it. Delays in the payment process are common. It's important for you or someone you designate to be responsible for making sure participants receive their incentives in a timely manner.

For remote tests, we typically send incentives via email. Usually we offer gift cards that can be redeemed online, but occasionally our clients prefer that we send them a check or other items.

In the United States, the Internal Revenue Service requires all organizations that provide $600 or more to individuals to report that fact on a 1099 form. The requirement is the same whether the incentive is cash, a check, a gift card, or a gift check. This requirement creates extra paperwork for the organization. Consequently, most organizations restrict incentives to less than $600 for any individual in one calendar year. If your organization conducts many studies, you may be required to keep records and make sure you do not go over the limit.

7.7.2 **Ending the session**

If your organization keeps a database of potential participants, this is a good time to ask participants if they want to be contacted for future activities. If so, you can provide them with a business card or a form to fill out so that they can be added to the database. You may also invite them to tell their friends or co-workers to do the same.

In a local test, ask participants if they know how to get back to their cars or to local transportation and if they need further directions. (See chapter 10 for ending sessions with participants with special needs.) In a remote test, walk them through the steps for ending the web conference and closing out any desktop sharing applications.

Your last action is to thank them once more for spending their valuable time to attend the session.

Finally, take a deep breath and a minute to feel good about yourself for the care and professionalism you used in the session. Don't be concerned that the session was not perfect. There has never been such a session. You did the best you could. And it's time to get ready for the next session!

INTERVIEW WITH AN EXPERIENCED MODERATOR

Have you ever had an issue with a participant after the session was over?

Yes, and here's what happened.

I ran a test with a report specialist. M had been a participant in my very first real test several months previously, and all of the report specialists in that group have a special place in my heart. I remembered him as a good participant—articulate and fitting the profile, the kind who puzzles things out and builds a mental model as he works with the product. I was glad to see him on the list.

He arrived on time and we recognized each other from the previous test. The session went well. He had less trouble than other participants, and the trouble he did have he was very articulate about. I was delighted.

M looks to be in his mid-50s, about 5 years older than I am. Neat, well groomed. At one point in the test I noticed that I found him attractive. Hmm, I thought. Isn't that interesting? Then I didn't think about it any more.

After the test, I escorted him to the front desk. It was after 5:00 P.M., so the receptionist was gone, but the security guy was at his post in the lobby.

M said that the testing location was better for him than his job. "Oh," I said, "Do you live nearby?" He said he lived about 30 miles away, but he works 15 miles farther away from our facility. Our office is in between. "Where do you live?" he asked me. "I live in W." He used to live in the next town, so we talked about the busy main street. It was a short, pleasant conversation, not unlike others I've had with participants after a test. We shook hands and said goodbye.

I give my card to all my participants so they know they can call or email me if they don't get the gift checks they expect in the mail. (And because I have a gazillion cards.)

It was a great test, and I felt great. My boss's boss had watched the test (remotely), and he sent me a very nice email afterwards. The product manager sent me a nice email about the test, also complimentary.

Then, around 1:00 pm the next day, I saw this message in my email:

K,

I wanted to express how much I enjoyed the
session yesterday. I would have to say that
your presence, professionalism, and enthusiasm
made what could have been a boring software
test rather enjoyable. You are very good at
keeping the process simple and at ease. I was
very happy to see you when you came to pick me
up at the front desk. I also enjoyed talking
after the session about my old 'hood.

Please feel free to contact me if you ever
need any help in the furniture world.

Thank you,

M

When I saw the email and the subject line ("Thank you"), I was puzzled but I thought, How nice!

It was very nice to hear about my professionalism . . . and it was nice to hear that my "presence" was a nice thing. I found it kind of funny that he called the tasks boring because I designed the tasks as well, but he doesn't know that.

As I read on, I got flummoxed. Oh, dear. I think he's flirting with me! Now what do I do?

He's a nice man, and not unattractive, and very professional himself, so this is flattering.

I had mentioned my husband last night, and he had said he was married, so why is he wanting to initiate contact? What is the thinking?

Did I give off a vibe? Oh, dear!

I told my boss that I thought my participant was flirting with me. He read the email and agreed that it was "a bit much." I asked him to keep an eye out for anything untoward I might have done as he watched the recording of the session. I asked my co-worker, who had watched half the session, if she thought I had been overly friendly. She said I had been just like I am with any other participant.

So, I was feeling embarrassed and wondering if I had something to be ashamed of. I felt sad because this was a good guy and a good participant in a profile that's hard to fill, and now I felt like maybe I should not have him as a participant again. I was upset because I meant to be a personable person, and somehow he had responded a little too strongly.

It occurred to me that my job in that room is paying attention to the participant. Being paid attention to for two hours is a very unusual occurrence for some people. Even if you go to therapy, you only get an hour! Bring a lonely guy into a room in the evening with a woman who pays attention to him, and perhaps there's a risk of misinterpretation.

Now, how could I respond in a human, professional way? Here's what I came up with and sent:

M,

Thank you for the kind words. I'm glad that you enjoyed participating in the test. You did a great job, and we did have a nice chat afterwards. It was a very good session.

Thank you again for your participation and for the compliments. I hope all goes well for you and your family up there in New Hampshire.

K

I never heard from him again.

Interacting in a remote test session

We are excited about the development of remote usability testing. We think that moving out of the laboratory environment is a positive step. Being able to include users from around the world expands the scope of testing to populations that were previously difficult or impossible to reach. Due to rapid advances in technology, remote testing has become popular very quickly. Many professionals like to moderate remote sessions because the sessions seem less stressful on both sides, once the technology is figured out. Perhaps the fact that neither moderators nor participants can see each other allows for a more relaxed posture and, generally, a more relaxed communication.

8.1 WHAT IS REMOTE TESTING?

As the name suggests, remote usability testing refers to testing sessions in which you and participant are not physically in the same place but are communicating via electronic technology.

8.1.1 Synchronous and asynchronous testing

There are two common types of remote testing: *synchronous* and *asynchronous*. In synchronous remote testing, the interactions between you and the participant are one-on-one and occur in real time. You and the participant communicate through shared technology.

In asynchronous remote testing, there is no moderator and participants work at their own pace. Participants' activities and feedback are recorded via special tools. Some asynchronous sessions involve survey questions triggered by user actions.

105

Because there is no interaction between you and the participant during asynchronous testing, we do not focus on it in this book. When we use the term "remote testing" in this chapter, we are referring to one-on-one sessions (synchronous testing). For more information about asynchronous testing, visit the web site *http://www. mangold-international.com*.

Although there are many variations on the setup of the remote testing environment, in this chapter we focus on what is in 2007 the most common setup. Moderators, participants, and visitors all have phones with either speakers or headsets. A phone collaboration product (e.g., a dial-in conference call number and password) is available through which all parties share voice communication. All parties use an application or web site through which they share their computer desktop and through which visitors can watch the screen activity. Control of the cursor can be passed back and forth between the moderator and the participant.

8.1.2 **The technology**

From the technical standpoint, the growing popularity of remote testing is a result of faster processors, increased disk storage, and the pervasiveness of the Internet. These technologies have made it possible to run the following key processes concurrently while maintaining acceptable performance:

- the product being tested (such as a web site or a software application)
- a sharing application (allowing parties to share control of the cursor)
- a recording application for storing the screen activity and voice communications

The ubiquity of broadband communications has given many people high-speed access to the Internet. Phone conferencing systems with access via a toll-free number allow multiple parties (test participants, moderators, note-takers, and clients) to participate in test sessions. (Note that not all U.S. vendors of phone conferencing provide access to users outside of North America.)

There are also collaboration and recording applications being developed specifically for remote usability testing, such as TechSmith's UserVue product (see *www.techsmith.com*) and Bolt, Peters, Inc.'s Ethnio (see *www.ethnio.com*). For a review of vendor and technology options, see the survey by Hawley and Dumas (2006).

8.1.3 **Advantages and disadvantages**

Compared with local testing, remote testing offers some advantages and disadvantages.

Advantages

- Expansion in the size of the sampling population. You are no longer restricted to the local area around your facility or to sites you can easily travel to. Anyone who has a phone, computer, and high-speed connection can participate in the study.
- You don't need a laboratory facility.
- It's generally easier to recruit participants because they don't have to travel to your location and you don't have to travel to theirs.
- You don't have to deal with late participants or late session starts caused by poor directions, parking problems, or traffic. You also save time by avoiding offering refreshments, escorting them to the test area, and so on. Some of these factors are very significant in big, busy cities.
- Because it takes so little effort to participate, participants often agree to participate at reduced or even no compensation.
- Participants can seem more comfortable during the session because they're in a familiar environment, usually at home or at their workplace.
- Participants have access to other computer applications or materials they routinely use, and the desktop and other personal settings are familiar.
- It is easier to involve participants who use assistive technology like a screen reader. In fact, it may be highly preferable to conduct such sessions remotely because the settings of these devices are often personalized.
- Finally, you sometimes gain some insight into participants' working environments that you would not get in the lab—noise, distractions, and so on. For example, you can get information about their systems, monitor sizes, computing power, and so on.

An added bonus is that the research to date suggests that remote sessions allow teams to find the same number of usability issues as face-to-face testing.

Disadvantages

- It is sometimes difficult to know what the participants are seeing on their screens (the participants can usually see the moderator's screen, but not vice versa.)
- The logistics of obtaining informed consent are more difficult.

- A small number of participants may have difficulty downloading or installing the desktop sharing software.
- Complex task instructions need to be available for viewing throughout the task (typically done using task cards in local tests). However, if you give the participants the task scenarios ahead of time by email or fax, they could look at them before you want them to.
- Because you can't control the participants' environments, participants can be interrupted or distracted during the session. We have had participants' children interrupt them, for example.
- Usually, you can't see the participants and their images are not recorded (unless they have a web camera on their end.) However, it's not clear yet how much of a disadvantage this is.

In our opinion, the disadvantages of remote testing are relatively minor given the benefits. We will explore each of the advantages and disadvantages in greater detail in this chapter.

WHAT THE RESEARCH SAYS

Does it matter if you cannot see participants' faces?

We are aware of only one study that is relevant to this issue. It was reported by Lesaigle and Biers in 2000. The study involved conducting a usability test and recording usability issues under three conditions:

- The evaluators could see only the screen activity of the participants.
- The evaluators could see the screen activity and hear what participants said (similar to most remote testing setups).
- The evaluators could see the screen activity, hear what participants said, and see the faces of participants (similar to lab-based testing).

The evaluators were then asked to create a list of usability issues from the recordings and to rate the severity of the issues they found.

The results showed no differences in the number of usability problems found in the three conditions. However, the evaluators who could see the faces of participants rated the same problems as more severe relative to those of evaluators in the other two conditions. There was no independent judgment of severity. Consequently, we do not know if seeing the faces made the higher severity ratings more accurate or less accurate. In addition, there was a low level of agreement among the usability professionals about which problems were the most severe.

The bottom line is that usability specialists saw something in participants' faces that was different from what they heard participants say and do on the screen. We don't know if what they saw allowed them to make better or worse judgments about the severity of problems. It does not tell us whether seeing faces in a remote test is a hindrance or a help.

Lesaigle, E. M., & Biers, D. W. (2000). Effect of type of information on real-time usability evaluation: Implications for remote usability testing. *Proceedings of the IEA 2000/HFES 2000 Congress, 6,* 585–588.

8.2 **PREPARING FOR THE SESSION**

In this section we discuss how to prepare and start a remote test session.

8.2.1 **Recruiting**

In a local session, almost all of the interaction takes place within one or two hours. Recruiting is often handled by a third-party recruiter and occurs before participants meet the moderator. Everything else is handled by the moderator—the moderator greets the participants, instructs them, monitors their task performance, and debriefs them— all within the allotted time.

In a remote test, on the other hand, there are several interactions between moderators and participants spread out over time. Moderators are more likely to recruit candidates themselves using email or the Internet, so their first contact is often electronic. Potential participants may come from a list of email addresses of customers or user group members, responses to an online survey, or people who monitor online classified advertising or message boards.

The general rules for recruiting (discussed in section 5.1) apply to remote testing. In addition to administering a set of screening questions to determine if participants qualify, there are some other logistical issues to address. By dealing with these issues right away, moderators can avoid wasting time setting up a session that can't be run. These issues include

- obtaining informed consent and agreement about confidentiality.
- determining what computing and communications equipment candidates have.
- establishing whether candidates are willing and able to download software to run the collaboration application.

8.2.2 **Obtaining informed consent and confidentiality**

As in any test situation, you need to obtain informed consent. That means preparing a consent form, sending it to participants, going over the points in the form, and having participants sign it. In remote testing, the challenge is getting the form to the participants ahead of time and getting the signed form back before the beginning of the test session. There are two common ways to deal with this. First is to use a paper form. You can fax or email the form to participants and have them fax the signed copy back to you. Figure 8.1 shows a sample consent form that can be used this way.

Second, you can create a consent form with an electronic signature, which you email participants to have them "sign" and send back. Figure 8.2 shows a sample electronic form. Using this form of signature is easier logistically than faxing, but you can use a form like this only if your organization will accept it as legally binding.

In addition to the consent form, your organization may require a nondisclosure or confidentiality form, which can be handled in the same way as the consent form.

It's best to get the signing of these forms done at least a day before the session. Otherwise, you may delay the test session while you wait for the signed form to arrive on your fax machine, or you may have to reschedule the session if, for example, participants don't have the authority to sign the confidentiality form.

Remember that you still have the responsibility to go over the informed consent with the participants even if they have signed it ahead of time.

8.2.3 **Determining the equipment candidates have**

Unlike lab-based testing, you can't rely on participants' having access to specific computing and communications equipment. As a result, one of the challenges of remote testing is establishing their technology setup. The specific equipment requirements, of course, depend on the test. Typically, the requirements when you're testing applications (as opposed to web sites) have to do with the six items that follow Figures 8.1 and 8.2.

Understanding Your Participation in This Study

Purpose
_____ is asking you to participate in a study of the _____ web site and its ease of use. By participating in this study, you will help us improve the design of the web site, making it easier to use.

Procedure
In this study, we will ask you to perform a series of tasks using the _____ web site. You will do this in your [*office/home*]. In order to make it possible to share your desktop, we will ask you to log in to a web site and download some sharing software. The moderator will be able to see your desktop but will not be able to see or change anything else on your computer. Afterward, we will ask you your impressions of the web site. The session will last about 60 minutes. We will use the information you give us, along with information from other people, to make recommendations for improving the web site.

Recording
We will record where you go on the web site (video) and the comments you make (audio). The recordings will be seen by only the _____ team that is analyzing the data and members of the _____ web site development team. After we complete the data analysis, we will send the audio/visual recordings to the client's design team, which will use them to improve the usability of the web site and not for any other purposes.

Confidentiality
Your name will not be identified with the data or recordings in any way. Also, only _____ employees who are working on the project will have access to the data we collect.

Risks
There are no foreseeable risks associated with this study.

Breaks
If you need a break at any time, just let us know.

Withdrawal
Your participation in this study is completely voluntary. You may withdraw from the study at any time without penalty.

Questions
If you have questions, you may ask now or at any time during the study. If you have questions after the study, you may call us at _____ or email us at _____.

By signing this form: You are indicating that you agree to the terms stated here and that you give us permission to use your voice, verbal statements, and videotaped image for the purpose of evaluating and improving the company's web site.

Signature: _____

Printed name: _____

Date: _____

■ **FIGURE 8.1** Sample remote testing Informed Consent form in word processor format.

Electronic Consent Agreement

Thank you for agreeing to participate in _____'s efforts to make our products easier to learn and use. This Consent Agreement describes the types of activities you may be participating in, the types of information we will be collecting during these activities, and how we will use this information.

Purpose: _____ is interested in learning more about the usability of its current products, and how to make its new products easier to learn and use.

Participation Methods: Your participation may include:
- Discussing your experience with the company's current products in an interview format
- Testing the usability of available products by performing tasks with them

Information Collected: Two types of information will be collected.
- During discussions and interviews, we will record the information that you share, including a voice recording. We may ask you to fill out a brief questionnaire regarding your use of _____ products, and/or other software better understand the context in which _____'s products are or may be used, as well as your comfort level products, to and familiarity with them and/or the products of others. Your name will not be included on the audio tape, the questionnaire, or any notes we take during discussions and interviews.
- If you participate in usability tests of current products or new prototypes, _____ will observe your screen activity as you perform tasks and record information about how you use the product and what you say about it as you are using it. _____ may also ask you to rate various aspects of a product's usability using a numerical or other scale. In addition, the company may electronically record your verbal statements and interactions with the product, including the mouse and keyboard. We will inform you before recording begins. Your name and other identifying information will not be included with these ratings, either on the notes we take during the test or with your verbal statements.

Your participation in this study is completely voluntary. You can withdraw from it at any time without penalty. If you have questions, you may ask them before or during the study. If you have questions after we're done, you can call us at _____ or email us at _____.

Please sign below to indicate your understanding and acceptance of this Consent Agreement.
    Thank you again for your participation.

Electronic Signature (*Filling in box serves as your electronic signature.*)

First and Last Name   Company/University   Email address

                                                                          (*Used for internal purposes only*)

Checking the "I agree" box and clicking on the Submit button, serves as your formal agreement to all of the terms listed above.

I Agree

| Submit | | Reset | | Date: _____ (*MM/DD/YYYY*) |

■ **FIGURE 8.2** Sample electronic consent form.

- Type and speed of the computer
- Availability of a phone with a speaker or a headset
- Availability of a high-speed Internet connection
- Type and version number of participants' Web browsers
- The firewall participants are using, if any
- A microphone that provides sound of an acceptable quality

Identifying participants' environment is not always easy. Many people don't know the specifications of the computers they use or the version number of their Web browsers. It may be necessary to talk with them and instruct them step-by-step on how to determine these.

If a project has special requirements, such as having a Windows operating system or a Firefox browser or avoiding systems with specific firewalls, allow yourself more time to set things up.

We suggest you determine your participants' setup as part of the screening process or while setting up the video collaboration software. It's best to get this done ahead of time to avoid taking up valuable time during the session.

You have a few options for ensuring that the audio connection is acceptable. Ideally, the phone-conferencing system's audio connection is integrated with the web-conferencing application. The benefit is that it allows easy management of all details of the communication and it offers the possibility for the audio and the screen portion of the session to be recorded together by the web-conferencing system. This ensures high-quality audio on the recording.

Web-conferencing products that do not have an integrated audio are still viable but require the moderator to manage the two systems separately and to record the audio via other means, for example, by placing a microphone beside the moderator's speaker phone. This setup will work, but the quality of the audio suffers. It works well enough for data analysis purposes but may not be good enough for video highlight clips.

8.2.4 **Establishing willingness and downloading software**

At the time of this writing, all of the publicly available video collaboration applications require some kind of download. The requirement to download software may disappear with time, especially for products that are developed just for usability testing. Some current products remove the downloaded application automatically when the session is over. A few products require only Flash (a browser plug-in that is often already installed on computers), but most applications require some additional software to be downloaded. The two non-Flash environments are a Java Runtime environment and an ActiveX environment. The Java environment usually requires a Java applet. This download can be done by participants following instructions in a dialog. The ActiveX environment, on the other hand, may require someone with administrator privileges to download an application. Some collaboration applications allow *either* a Java or an ActiveX environment, though they are often are more expensive. (See Hawley & Dumas, 2006, for an explanation of these alternatives.) We have had downloading problems in only a handful of the sessions we have run.

When participants are using their computers at work, there may be firewall issues, such as a block on downloading all or certain types of applications. For example, some government offices and health care organizations allow *no* downloads. Unfortunately, you never know when these situations are going to arise, so participants have to try to join a collaboration meeting before you can tell whether it will work. Be aware that it is not enough to have participants simply go to the collaboration meeting web site; the downloads are required for *joining* a meeting, not just getting to the site.

Fortunately, if you work through the process with participants, you will be successful most of the time. We have also found that when a download doesn't work, it is not worth having the candidate contact his or her IT department. Usually that just wastes everyone's time because the issue is policy-related rather than technical. It's best to find another participant.

8.3 INTERACTING DURING THE SESSION

In this section, we cover

- establishing what participants are seeing
- providing instructions on thinking aloud
- dealing with task scenarios
- avoiding dependencies between tasks
- managing visitors
- dealing with distractions at the participant's end

WHAT THE RESEARCH SAYS

Are remote and face-to-face tests comparable?

A study by Brush, Ames, and Davis (2004) compared the results of a remote and a local test. There were 20 participants—12 were remote and 8 were local. The participants were all urban planners. The product they tested was an urban planning simulation tool. In the remote sessions, the moderator could see the screen activity and hear participants but not see them. In the local sessions, the moderator sat beside the participants in the test room. The same moderator ran all 20 sessions.

An interesting aspect of the study was that after the sessions, 4 participants in each group were asked to participate in a second study using the opposite condition—that is, 4 participants were in a remote session first and then in a second local session, while 4 participants were in the local session first, then in a remote session. This arrangement allowed the participants to directly compare the two experiences.

The analysis of the data from the first 20 sessions showed that there were no differences between the remote and local groups in the number of problems uncovered or in the types of problems uncovered. This finding is important because it indicates that remote tests have just as much diagnostic power as local tests, even though the moderator cannot see the participants. There were also no differences in the severity of the problems identified.

Of the 8 participants who were in both conditions, 7 felt that it was more convenient to be in the remote session. Four participants preferred being in the remote session while the other 4 said that they had no preference. Three participants felt that it was easier to concentrate on the tasks in the local setting, while only 1 participant felt that it was easier to concentrate in the remote setting.

From the moderator's perspective, the investigators reported the following results:

- It took more effort to prepare for and set up the remote sessions.
- Recruiting remote participants was easier.
- It was more difficult to deal with software crashes and network failures in remote sessions, although this is less of a problem than in the past.
- Remote participants were interrupted a total of nine times. These interruptions had no impact on this test, but it could have if task time had been measured.
- There was no difference in the average time to run the two types of sessions.

On the whole, remote and local tests yield remarkably similar results.

Brush, A., Ames, M., & Davis, J. (2004). A comparison of synchronous remote and local usability studies for an expert interface. *CHI 2004*, 1179–1182.

8.3.1 **Establishing what participants see on their screen**

This is a surprisingly difficult problem to deal with. You cannot assume that when you share your desktop or an application with participants that they are seeing what you want them to see. Sometimes the difference is a minor one, such as the bottom or top of the screen being cut off. Other times it's much more significant, such as different screen resolutions or the participant's window is not maximized. Asking the simple question "Tell me what you're seeing on your screen" does not solve this problem. Here are some simple questions you can ask to get a better idea about what they're seeing:

- What are you seeing at the top (and bottom) of your screen? If they say that they see a menu, you might ask them to read it to you.
- Are you seeing a scroll bar on the right-hand side (or the bottom) of the screen? Sometimes they should be seeing one, but at other times the bar may indicate that they are only seeing part of the screen you're seeing.
- What does the maximize button on the top of your screen look like? If it shows two boxes, the window is maximized.

You also can display a screen shot on a slide that shows them what they should be seeing. Video 5 on this book's web site illustrates this issue.

8.3.2 **Providing instructions on thinking aloud**

As discussed in chapter 5, moderators have different ways of providing instructions on thinking aloud. Some moderators model the think-aloud process for participants by manipulating some object while speaking, for example, replacing the staples in a stapler. Some moderators have participants practice by manipulating some object while thinking aloud. This type of training with physical objects obviously won't work in a remote session.

One solution to this problem is to have the moderator display an unrelated web site or application on the screen and share it with participants. The moderator then performs a task while thinking out loud as participants watch and listen. They can then do a different task with the same site. We have noticed a tendency for participants to mimic the moderator's comments, however, so be sure you model different types of comments (e.g., something positive, something negative, something unexpected, something related to workflow).

There is no evidence that a demonstration of thinking aloud increases the likelihood that participants will talk more. In addition, we don't know if participants in a remote session talk more or less than participants in a local session. Our impression is that most moderators do not model or have participants practice thinking aloud. Until we see some research on this topic, we have no basis for knowing which practice is best.

8.3.3 Making the task scenarios available to participants

In a face-to-face test, the task scenarios typically are printed on sheets of paper or on task cards. Participants read the scenarios and proceed to attempt the tasks. The advantages of this procedure are that participants can only see one scenario at a time and they have it in front of them throughout the task. If there's any question about what the task is, they have the card to refer to.

In a remote test, if the scenarios are short and easy to remember, they can be typed into a word processing or presentation file. Then each scenario is shown to participants, and then the window is minimized while the participant is working on the task. Alternatively, the moderator can give the tasks verbally. We have found that trying to use instant messaging or an electronic whiteboard usually takes up too much screen real estate.

If the scenarios are long, have multiple steps, or are complex, it may not be possible for participants to remember them for the duration of the task. In those situations, it's too cumbersome or intrusive to keep displaying the scenarios repeatedly to participants by minimizing and maximizing the window. A common solution to this problem is to email or fax the scenarios to participants before the session starts. However, this causes you to lose control of how and when participants see the scenarios, which could have an impact on their performance of the tasks and thus affect the test results. You can ask them not to look at the scenarios before the session and not to look ahead at future scenarios during the session, but you can't be sure they will comply. In our experience, asking them not to look ahead usually works well, but there's no guarantee that they have not previewed what you sent them.

Additionally, there are some tests during which a task is inserted in the midst of a scenario. In those cases, you may want to present the inserted task verbally or on a slide at the appropriate time. You can make this seem less like a trick and more like a spontaneous event by framing it as unplanned. For example, you might say, "What would happen if you tried to . . ."

8.3.4 **Avoiding dependencies between tasks**

In some tests, the ability to complete a task depends on having successfully completed an earlier task. For example, completing a calculation in a table or spreadsheet may depend on the correct data having been placed there in an earlier task. It's a good practice to avoid these dependencies but sometimes there is no choice.

In a local test, there are three ways for a moderator to deal with dependencies if participants can't complete the earlier task: (1) complete the earlier task for them, perhaps asking them to look away; (2) have a second instance of the file or application prepared that's in the appropriate state, and then open it for the later task; or (3) assist them until they get it right. (See chapter 6 for further discussion of these options.) The first two strategies are cumbersome to implement in a remote test because you have to take back control of the cursor. To complete the earlier task without participants seeing it, you may have to turn off screen-sharing, complete the task, then turn it on again. To open a file with the prepared solution in it, you have to go to where the file is located to open it. In both of these cases, your actions may call attention to the fact that the participant could not complete the earlier task.

8.3.5 **Managing visitors during the session**

People whom you or some other team member invite to observe the session may be in the room with you or may join the session remotely through the collaboration tool. When the visitors are in the room with you, you can control what they say and do just as if they were in the control room at a local test.

When the visitors join through a collaboration tool, however, there are a few issues to deal with.

Eliminating background noise on the visitors' phone. Sometimes it's noise from traffic outside; other times someone comes in the visitor's office and starts a conversation. Some phone collaboration products allow the session leader to mute visitors' phones; others do not. If you can't control the muting, it's important for you to tell the visitors to do it themselves.

Preventing visitors from interrupting the session. Muting visitors' phones also prevents them from talking during the session when you don't want them to. But this may not prevent them from un-muting themselves. In our experience, the majority of visitors will keep silent. But occasionally someone does interrupt. You need to be explicit to visitors about what they can and can't do during the session. It's a good idea to give them instructions in an email prior to the session.

Allowing the participant to see the list of attendees. Sometimes it doesn't matter if participants see the list of attendees to the conference. But in some situations, you may not want participants to know how many people are observing or who they are. In that case, you should hide the list of attendees. Most collaboration tools display a list of the attendees, often on the right-hand side of the session leader's screen. Some tools also show this list to all parties. There's usually an option to turn this display off for everyone but the session leader. Also, keep in mind that if you share your entire desktop with participants, they will be able to see the list on your screen unless you hide it or turn it off.

8.3.6 **Dealing with distractions at the participants' end**

One of the downsides of having participants work in their own environments is that you have less control over distractions and interruptions. For example, a common interruption at the participant's end is a phone call. As with phone calls in a local test, try not to allow them unless there is an emergency. Ask participants to turn off their cell phones and beepers at the start of the session. However, this is not always effective. Sometimes you have no choice, such as when the participant puts you on hold without asking. Our view is that occasional phone interruptions don't usually affect the validity of the diagnostic remote sessions.

Colleagues or family members also may walk into participants' offices or cubicles or home work areas. Although we always tell participants ahead of time that they will need an uninterrupted hour for the session, this doesn't always work. Some interruptions are minor; others are not.

We were once in this situation while testing customers of a product. The participants' managers knew about the test and were curious about the new product. Consequently, one or two stayed with participants during the remote usability sessions. Several times, they interrupted the test sessions because they wanted to see what all the interest was about. They wanted to stay a while and "just watch." Unfortunately, this situation could have put the participants on edge, so after a few sessions, we politely asked the managers to leave.

In our experience, it's a rare manager who can really "just watch." Overall, having a manager present during the testing session in a remote as well as traditional lab environment may put participants in an awkward position, which would impact the outcome of the testing and bias the data.

So, sometimes you need to try to get the visitor to leave. You can be straightforward and say that you want to get participants' reactions to the product as they work alone, without interruption or help. This approach should work most of the time. If the visitor persists, another strategy is to offer to include the visitor as a test participant in a separate session. In fact, in our experience, some visitors, particularly managers, want to be in the test themselves and ask to be included. Although you may decide not to include their data (if they don't match the user profile, for example), running a session with the manager is often a good public relations move and worth the time it takes.

INTERVIEW WITH AN EXPERIENCED MODERATOR

Remote testing comes with some unique challenges because you can't see your participants. They are in their workplaces rather than yours and as a result you have less control. How do you handle this?

On one particular occasion, I had one of our key customers scheduled for a session. When "Todd" joined the call, to my surprise, he was not alone. Someone else was on the line with him. "I thought Sue would be really interested in joining us, so I invited her along. I hope that is OK," he said. Ack! My mind raced. Who is Sue? Why did he think he could invite her? What should I do here? I determined that the best thing to do first would be to determine Sue's job role. A quick assessment of her job function and seniority might help me determine what to do next. It turned out that Sue was a co-worker of Todd's and that she too was a financial analyst. We still had some openings that week, so I explained to Sue that I was very pleased that she wanted to take part but that the sessions were designed to be one-on-one so that I could understand each individual's experiences in depth. I asked if she would be willing to take one of the open slots later in the week so I could learn about her specific experiences with the prototype.

Luckily, Sue was more than happy to take part in a session later that day, so it was a happy ending.

This situation is not uncommon when conducting remote testing. In other situations, I have been surprised to find an executive or other key stakeholder on the call. This is a very sensitive situation, and you need to remember the priorities: (1) the customer/participant, (2) your company, (3) the data.

If you can't remove the surprise visitor from the session without causing harm, then you should continue the session with both participants. In this situation, I have asked one of the participants to take the lead in completing the tasks, and then asked for comments from the other person at the end of each task. In the case of an early formative test, you can still get some good data. In the case of a summative test, you will likely need to sacrifice the data for the sake of the customer and your company.

8.3.7 **Making a connection despite the physical distance**

You will recall from the discussion in chapter 3 how much connecting with participants depends on nonverbal cues, such as eye contact and body language. In a remote test, you have only your voice and what you say to make and keep a connection. Be sure to use varied intonation rather than a flat, dull tone of voice.

Also, pay very close attention to what the participant is saying and doing. Our recommendations:

- Listen for changes in tone of voice, hesitations, and sighs, which may signal frustration.
- Keep in mind that silence can indicate many things in a remote session. For example, the participant may be squinting at the screen or have a puzzled look. After a silent period, ask what is happening.
- Make a special point in between tasks of providing positive comments, such as "This is really helpful."

We heard of one company that avoids this issue altogether by sending participants a web camera ahead of time so that they can see participants' faces as they work. Afterward, participants keep the webcam as their incentive.

Moderator–participant arrangements

If you visit a sample of usability labs, you will see at least one striking difference among them. Some organizations normally have moderators sit in the test room with participants, and others have moderators sit in a separate room and talk with participants over an intercom. These two arrangements (in versus out of the room) reflect differences in philosophy that started more than twenty years ago. Today, there are many more possible arrangements in a lab setting and when you are not using a lab.

This chapter explores a number of arrangements and discusses their advantages and disadvantages. Our goal is to present the options fairly so you can make your own decision because we've found that this issue can be very controversial, perhaps the most controversial issue in this book.

There are strong opinions about the effects of moderator–participant arrangements on the validity of testing. Experienced moderators disagree about the value of physical closeness (and distance) from participants and how it does or doesn't affect the amount and the quality of data collected in a test session. In addition, some believe that being physically separated in adjacent rooms and communicating over an intercom affects the power relationship between the moderator and participant, that it overemphasizes the Leader role over the Gracious Host role. We find it remarkable that moderator–participant arrangements have created strong opinions but there has been no research on their impact on the testing process.

This chapter covers

- a brief history of arrangements.
- various types of arrangements.

- advantages and disadvantages of the arrangements.
- situations that dictate which arrangement to use.
- considerations for the practitioner.

9.1 **A BIT OF HISTORY**

Most of the books that describe testing from a didactic perspective don't help practitioners understand whether they should sit with the test participant or in a separate room. Dumas and Redish (1999) assume that the moderator and participant are normally separated: "In most tests in which there are separate observation and test rooms, the participant works alone in the test room" (p. 295). Because one of the authors of that book (JD) is also an author of this one, he can tell you that in 1991—at the time the first edition was being written— all of the testing he had done was diagnostic and as a moderator he was always separated from the participant. He assumed that this method was the one that everyone used. But he was wrong.

In its early days, usability testing was considered a type of research experiment. The first description of testing, Roberts and Moran (1982), described it as similar to a research study. With that view-point, the emphasis was on data integrity. The moderator's role was to act like an experimenter maintaining an emotional and, whenever possible, a physical distance from participants so as not to influence them. As late as 1993, the term "experimenter" was used to describe the moderator in a usability test (Nielsen, 1993). Rubin (1994) did describe physical separation as creating an impersonal environment. "This is sometimes referred to as the 'guinea pig' syndrome, with the participant feeling overly self-conscious during the test. The effect can be exacerbated by the type of intercom system used, some of which make the test monitor's instructions sound like the 'voice of God'" (p. 57). He seemed to favor being physically beside the participant without saying so explicitly.

Snyder (2003), however, is very clear about her view that the moderator needs to be in the room with the participant when a paper prototype is being tested but does not generalize that requirement to other types of products.

There is one study that is remotely related to this issue. Barker and Biers (1994) conducted an experiment in which they varied whether there was a one-way mirror or cameras in the test room. They found that the presence of the equipment did not affect the participants' performance or ratings of usability of the product. This finding showed that the testing environment did not affect the diagnostic value of testing but it did not test directly whether participants are intimidated by that environment.

9.2 PHYSICAL ARRANGEMENT

Since the mid-1990s, there has been a blurring of the boundaries among usability evaluation methods. Testers have explored many options to find the right mix of procedures that will maximize the quality of data as well as the participants' experience.

9.2.1 The moderator in the test room

First, there are several arrangements where the moderator and the participant are in the same room. Most typically, the moderator sits next to and slightly behind the participant. Often the seating is slightly angled to create a natural posture for conversation. When testing software, web sites, or electronic prototypes of physical devices, the moderator needs to be able to see the computer display on which the participant is working.

Sometimes, after starting participants on a complex or lengthy task, the moderator moves away from the table and sits on the other side

■ **FIGURE 9.1** The moderator sits to the side and slightly behind the participant.

of the room until the participants say they have completed the task. In this case, there is a note-taker in another room.

Finally, the moderator can sit across a table from the participant and watch what the participant is doing on a second monitor. This variation is useful for situations where the subject of the test is sensitive and the moderator does not want to appear to be looking over the participant's shoulder.

9.2.2 **The moderator not in the test room**

Sometimes the moderator starts the task and then finds an excuse to leave the room, allowing participants to work on their own. The note-taker follows the task and is able to confer on whether the participant's behavior seemed to change when the moderator left. This may be useful for participants who seem particularly shy or have started to develop a dependency on the moderator for cues.

And, of course, there is the arrangement when the moderator leaves the room after the initial briefing and stays in a separate room until the tasks are complete, returning to the test room for the final activities (questionnaire, interview, etc.).

9.3 **BELIEFS ABOUT ARRANGEMENTS**

The question of whether the moderator should be in the same physical space as the participant is based on different views about (1) maintaining the integrity of the data, (2) the importance of the power relationship with the participant, and (3) making the participant feel comfortable while allowing closer observation of what he or she is doing.

9.3.1 **Physical separation**

Moderators who choose to be physically separated, when it's appropriate, believe that the separation makes it less likely that their interactions with the participant will affect the quality of the data. They believe that participants are less likely to ask for help or change the way they interact with the product if they are not in the room. In addition, moderators, especially beginners, are less likely to be drawn into conversations with the participant, to give hints or cues, or to send signals of approval or disapproval through their body language.

People who hold this view agree that some participants are uncomfortable when they are alone, but that most of them quickly get over that feeling and, after a few minutes, it doesn't matter to them whether they're alone in the room. Consequently, those moderators believe that the data is less likely to be biased when they're separated from participants and that the increased discomfort that some participants might initially feel does not justify the risk of influencing the data by remaining in the room during the test.

Those moderators also believe that there are additional advantages to being in a separate room. For example, they can interact with developers and visitors to explain what they're seeing. They can ask developers important questions that clarify what participants are doing and whether the product is working properly. The separation also makes it easier for them to take measurements such as task times and to log events, eliminating the need for a second person to conduct those activities.

Finally, some of these moderators report that being in a separate room is less stressful for them. They don't have to worry about their body language or being drawn into a discussion with the participant. These moderators interact over the communication system, but they don't feel that they are "on stage" all the time.

9.3.2 **Physical proximity**

Moderators who choose to be in the room with participants believe that participants' discomfort with being left alone in the test room can be substantial and can affect their performance. The idea is that when sitting alone in a room (especially when asked to think aloud), participants are more likely to feel nervous and self-conscious and, as a result, persist in unproductive paths when working with the product. In essence, these moderators believe that physical separation leads to emotional separation. The feeling of isolation can be worsened by poor microphone techniques on the part of the moderator.

Some of those moderators feel that subjecting participants to discomfort borders on unethical because it's unnecessary. By sitting with the participants, moderators can provide emotional support that makes them more comfortable in the testing environment.

Those people believe that experienced moderators can avoid being drawn into unnecessary interactions, giving cues to participants, and being biased. Furthermore, they point out that the presence of visitors in an *observation* room can be distracting to moderators, drawing them into conversations that take them away from being effective at moderating.

Another advantage to being beside participants is that it is easier to see what they are doing.

A final point is that moderators who believe in physical closeness talk about the contrast between equality and power. They believe that the best arrangement for participants and for the quality of the data is one that fosters equality of roles between moderators and participants. At least in North American culture, the assumption is that power equality leads to less stress on participants and allows them to act more "naturally."

9.4 **CHOICE OF ARRANGEMENT**

Up to this point we've been talking about a moderator's or testing group's default method for interacting in cases where there's an option for being in the room or separated from participants. But some testing situations dictate which arrangement is best. There are three factors that determine whether the moderator should be beside or separated from participants:

- objective of the test
- interactivity of the product
- characteristics of the participants

9.4.1 **Being physically close to participants**

Some tests clearly require the moderator and participants to be in the same place.

The focus is on exploring design alternatives or the conceptual model that underlies the user interface. These kinds of test usually are held early in product development. Task performance is less important than understanding participants' thoughts and perceptions. Consequently, there is continual conversation between you and participants and you probe frequently to understand the mental model participants are forming of the product interface. There often is much pointing at the screen to focus on details. These tests are best done with you sitting beside the participant.

A lack of interactivity in the product requires help from the moderator. These tests also happen early in development; examples are tests of paper prototypes, static screenshots, and user interfaces with only some of the functions active or only the correct path to task success active. In these situations, you're likely to be responsible for communicating how the interface is *intended* to work or, when participants have selected the correct course of action but nothing happens, explaining why the prototype is not showing the result.

Snyder (2003) makes a persuasive case for having the moderator sit with and interact frequently with participants while evaluating a paper prototype. In her extensive experience, she's found that thinking aloud does not work well when the product is being manipulated by a "computer," that is, the person who makes the paper prototype interactive. Participants tend to talk to the person who is the computer rather than think aloud or talk to the moderator. To avoid this conflict, the moderator asks participants questions and keeps their focus on the prototype. If there are visitors at the table, the moderator has to manage interactions with them as well.

Sometimes with a static screen-based prototype, such as a Visio™ or PowerPoint™ file, the moderator controls the mouse and moves through the screens. Participants are then asked what they *expect* will happen next because it can't be demonstrated. These additional requirements clearly require the moderator's presence.

There are at least two types of participant who are likely to require the moderator to sit with them.

1. Participants who require special help using the product or the equipment—examples are people with disabilities, young children, the elderly, or people who are completely inexperienced with the equipment they will be using.
2. Participants who may be upset or disruptive if left alone, such as some teenagers (Hanna, Risden, & Alexander, 1997).

We have more to say about interacting with these populations in chapter 10.

9.4.2 **Being physically separated from participants**

Some tests yield the desired data better when the moderator and participants are not in the same place.

Focus is on quantitative measurements of usability. As a group, these measurement tests are "cleaner" when the moderator is outside of the room.

Focus is on task times and/or other measurements. In some situations, such as benchmarking studies, it's critical to collect time-on-task data. Test protocols usually ask participants to avoid verbalizations, which can lengthen task times. Furthermore, participants may be asked to work at a steady pace but not to rush through tasks.

Assistance is prohibited. Moderators are sometimes told to avoid giving assistance to participants under all circumstances. This can be difficult to do from within the room, especially if there are many task failures. During those tests, interactions between you and the participants are minimized, especially interactions that are not part of the measurement plan. Probing, diagnostic discussions, and even encouragement are generally not allowed. The goal is to make sure that whatever interactions take place are not a source of bias.

Measurement tests are best done with physical separation. The need to record data and minimize interaction is facilitated by the separation. In these tests, any emotional connection with participants can be lost if you're not careful.

You are responsible for recording complex data. Sometimes you're extremely busy recording detailed measurements. For example, there may be several steps within each task to measure and there may be specific time limits for tasks, or paths to task success or failure may need to be recorded. If there isn't another person to help record the data, it can be less intrusive if you perform these activities from another room.

9.5 **CONSIDERATIONS FOR THE PRACTITIONER**

There are a number of tests in which you have several options about the moderator–participant arrangement. It is best to base your decision on the objectives of the test, the interactivity of the product, and the characteristics of the participants. We believe that several arrangements can work and that until we get some research data, you should do what you think is best for you and the participant and follow our guidelines in chapter 3 for making and maintaining a connection with participants.

INTERVIEW WITH AN EXPERIENCED MODERATOR

When you have a choice, do you sit with the participant or in a separate room?

When we built our usability labs, we went to a fair amount of trouble to set them up so that the facilitator was not in the room with the participant (and by the way, we called them experimenters and subjects back then). We tried several kinds of microphone arrangements and video camera placements so that we could see and hear what was going on and communicate effectively with the participant. In general, I would say the overall effect from the participant's perspective was cold and unnerving, especially if we didn't get the volume right and the "voice of God" boomed into the room. As a facilitator, it was very hard to see everything that was going on and to accurately gauge the intent of the participants' actions and their states of mind.

For a number of reasons, we began conducting evaluations in the same room as the participant. One reason was that we started to use low-fidelity prototypes that required us to be in the room to simulate the system. Another reason was that a new colleague joining the group had always conducted sessions while in the room and continued to do so after joining us! Another big reason was that we began to do a great deal more ethnographic work, where we went into the participants' environment, and most of these environments were conspicuously missing one-way mirrors and intercoms.

Several benefits became clear when we were in the room. First, interaction with the participant was much more natural, having more of a feel of co-discovery than a test. We were much more able to read body language, see where the participant was looking, and so on. These nuances are important to understanding what's going on, why a person may be misunderstanding something, and most important, what might need to be done to fix the problem.

There was no single "aha!" experience or decree that "thou shall be in the room." Over time we simply abandoned the clunky intercom and camera arrangement and began doing all of our testing in the room with the participant. I think that we get more out of each session, and this is borne out by the anecdotal finding that observers outside the room, monitoring remotely, always defer to the facilitator in the room when discussing what had actually happened.

Chapter

10

Interacting with diverse populations

As usability professionals, we have a responsibility to design and evaluate products for the widest possible market. User interfaces should be intuitive for all their users, not just the computer-literate and able-bodied ones. Moderating usability tests with some populations, however, requires additional knowledge, training, and preparation. Even so, moderating a session that includes assistive technologies, interpreters, or multiple modes of communication can be a very rewarding experience.

Although there are many guidelines in this chapter, we don't want to give you the impression that interacting with special populations is complicated. The many reminders are simply an extension of being courteous to others, with a few specifics that might not be common knowledge. Being comfortable with diverse populations becomes easier with familiarity, and even if interacting with these populations is new to you, you're not likely to be dealing with more than one of them at a time. Consequently, you can use this chapter as a reference.

This chapter provides guidelines for interacting with a several populations:

- People with physical disabilities
- The elderly
- People with cognitive disabilities or low literacy skills
- Children and teens
- People from cultures different from the moderator's

Some of the guidelines come from our own experience and others come from people who have interacted with these populations and have shared their experiences in the literature, particularly Hass (2004), Hartman (2005), and Henry (2007). We have restricted our

Table 10.1 Useful Web Sites with Information about Special Populations

What is on the site	URL
Information about service animals	*http://www.people.howstuffworks.com/guide-dog.htm*
How to find interpreters	*http://www.rid.org*
ADA checklist of ways to make facilities accessible	*http://www.usdoj.gov/crt/ada/votingck.htm*
American Foundation for the Blind	*http://www.afb.org/default.asp*
National Association of the Deaf	*http://www.nad.org*
How to locate a CART provider	*http://www.cartinfo.org/locate.html*
How to locate assistive listening devices	*http://www.hearinglossweb.com/res/ald/ald.htm*
Procedures for developing materials for low-literacy readers	*http://www.cancer.gov/aboutnci/oc/clear-and-simple*
Braille embossing services	*http://www.universalmediaservices.org/BrailleProduction/Embossing.htm*

discussion to what we believe are the most important issues to consider in moderating sessions with these participants.

Table 10.1 contains a list of useful web sites about many of the issues we discuss here.

10.1 GENERAL GUIDELINES

The guidelines in this section apply to all of the groups discussed in the chapter. Later sections of this chapter provide more details about distinct populations.

Treat each participant as an individual. Every human being is unique. Not all "blind" people are sightless; not all 80-year-olds have poor memories. You can't stereotype them any more than you can stereotype the general population. Part of a moderator's responsibility is to assess each participant's ability to understand

and perform during a session. Your observations from the time you screen them as candidates, through meeting them as participants (in person, over the phone, or via the Internet), up to the time they start working on tasks will help you to decide whether they have the knowledge and skills to evaluate the usability of the product or if they need additional help to do so. This assessment is especially important. You may have to prepare more thoroughly for, or work harder in, sessions with people who have special needs, but it's worth the effort.

Ask for feedback. At the end of a pilot session and after the first few test sessions with participants from a new population, it's helpful to ask, "We will be conducting more sessions with participants who are [blind, elderly, etc.]; what would you recommend that we do to make them comfortable during sessions similar to yours?"

Ask before you help. In many cases, people with limitations may seem to be having difficulty but would prefer to complete a task themselves. If a participant declines an offer for help, don't feel offended. And it's especially important to respect the wishes of helpers and aides.

Learn about the population. Until you have extensive experience with a population, it will be very helpful to contact relevant national and local groups that support the population. Many have web sites that contain useful reading materials. These organizations are more than willing to provide advice when you contact them. As an example, schools and colleges for people with limitations are often an excellent resource. They have a long-term outlook and maintain relevant equipment and facilities.

Make instructions and explanations simple and clear. In addition to the clarity of language and logical presentation of instructions, it might be helpful to moderate your speaking pace or tone. You can't expect everyone to take in and process information at the same rate. For a variety of reasons, participants may have difficulty perceiving or understanding what you're asking of them. You may have to explain more terms or concepts. They may be easily distracted, especially when they're interrupted or have to deal with crosstalk. Go slowly and watch for signs of confusion.

Be sure participants understand and sign the informed consent form. As mentioned in chapter 5, obtaining truly informed consent is one of the moderator's most important responsibilities. Each group we discuss in this chapter has limitations you may have to accommodate to obtain truly informed consent. For example, you may have to translate the consent forms into embossed Braille, or explain participants' rights to both parents and children, or read the form out loud to participants with lower literacy skills.

Consider the ability to handle concurrent thinking aloud. Several studies show that some groups of participants have difficulty performing tasks while thinking out loud. All humans have a limited cognitive capacity (Anderson, 2000), and some populations have additional limitations. There is a growing literature on the "reactivity" of thinking aloud, which refers to situations where thinking aloud and working on a task negatively interact. Trying to perform both activities at the same time changes both the think-aloud and the task performance. For example, Branch (2000) studied reactivity in middle school children who were trying to find information on a CD. The children with the least computer experience performed the information-seeking tasks with the least success and their thinking aloud tended to be limited to descriptive "play-by-play" statements rather than evaluative statements. Birru and colleagues (2004) found that people with low literacy skills had difficulty thinking aloud while they worked. Although we are not aware of similar research with the elderly, many elderly and people for whom the test language is not their first language may well have difficulty with concurrent thinking aloud. For the populations we discuss in this chapter, it may be more effective to have participants do *retrospective* thinking aloud, which occurs after each task or after all of the tasks are completed.

Expect tasks to take longer and make sessions shorter. Testing sessions with people who have limitations will not be as efficient as those with people who do not have those limitations. Your instructions and explanations will take longer and often the participants will work more slowly. Our guideline is to expect sessions to take at least 25 percent longer. In addition, a two-hour session that participants without those limitations can get through is likely to be too long for many participants in special groups.

Provide adequate breaks. Special populations often require more and longer breaks than others. Some need frequent breaks to maintain their attention, and some may have health concerns that require frequent restroom breaks. Taking a break every 15 minutes is common.

Be prepared to provide escorts. Many participants from special populations can't drive cars. They may need escorts to and from the nearest public transportation facility or from a taxi cab. They may also need an escort to the restroom and help obtaining a taxi after the session. In these situations, you will need extra staff to serve as escorts or, if you're the escort, you will need to leave extra time before and after each session.

Expect guests. Many people in special populations have aides who will come with them to a session. In the case of children, the aides are their parents. You need to provide a comfortable place for the aides to wait and something for them to read or do. If you expect service animals (such as guide dogs) to be present, anticipate their needs as well (see section 10.2.2). In some cases, the aide may need to be in the room with the participant during the session. Never interfere with or hinder the duties of an aide. There may also be legal stipulations dictating access. For example, just as it is illegal for a restaurant to refuse service to a blind customer with a service animal, it is illegal to ask a blind participant to leave his or her animal or aide outside.

Refer to people and their abilities and disabilities in a respectful way. It is, of course, important to address all participants respectfully, which includes being sensitive to the limitations of participants when it is necessary to mention them. Our experience is that, if anything, you will try too hard to be sensitive. But just because there is a euphemism for almost every condition or situation doesn't mean you need to use it. Furthermore, individuals vary greatly in their sensitivity. If you use language that doesn't reflect how an individual prefers to refer to himself or herself or his or her disability, apologize, ask for the preferred term, and move on. Some blind people may think you are talking down to them if you refer to them as "visually impaired" or "sightless," and others won't care which term you use. Henry (2007) suggests always putting the person first, such as "a man who is blind" rather than "a blind man." Similarly, we say, "a man who uses a wheelchair" rather than "a wheelchair-bound man."

If you work with diverse groups, it's likely that you'll inadvertently use a term that some participants object to. When that happens, apologize briefly, and ask for the specific terms they would prefer. For example, if a participant objects to being referred to as "sightless," you might ask, "I'm sorry, which term do you prefer?" Then get back to the activity you were performing. You are likely to feel bad no matter how innocent the intention, but you must set aside those feelings to continue to moderate effectively. Furthermore, trying to explain yourself with additional elaborations usually gets you into more trouble. It's a good idea to talk about these situations with colleagues after the session, and we urge you not to let it make you hesitant with the next participant.

Monitor your feelings. Even though most moderators are people-oriented, it might be uncomfortable to work with diverse populations at first. Until you have a lot of experience with special populations, it may be quite a challenge to interact with them—to know how to address them, to know how to help them without insulting them, to know how to prompt and probe (*Can she do more? Is that all he can do? Should I push for more on this task or move on?*). Be aware that you may be less willing to push participants to continue working, or you may be more likely to provide assistance than you normally would. As a moderator, you must decide what is appropriate and trust your intuition. The important thing is that you make those decisions consciously rather than unconsciously.

10.2 **PEOPLE WITH PHYSICAL DISABILITIES**

Fortunately, a number of usability practitioners have extensive experience moderating tests with participants who have physical disabilities. Good sources for practical guidance are Hass (2004), Hartman

■ **FIGURE 10.1** Temporary barriers to navigation for some disabled participants.

(2005), and Henry (2007). We start with guidelines that apply to most people with physical limitations, then present some guidelines that apply to people who are blind and people who are deaf. Section 10.4 discusses participants with learning disabilities and low literacy.

10.2.1 **Interacting with the physically disabled**

Provide accessible facilities and restrooms. This guideline seems obvious, but there is more to it than appears on the surface. Providing "access" means that the entire path from the street to the testing facility must be accessible. Look for appropriate curb cuts, elevators that are wide enough for wheelchairs and equipped with controls for blind participants, and so on. Before the first day of testing, walk the path to and through your facility and look at it from the participants' perspective.

Remove obstacles such as chairs and wastebaskets and wet-floor signs that might get in the way of wheelchairs or blind participants. Be sure to have enough space in your test room for flexible furniture arrangements. Wheelchairs and participants with canes or service dogs need lots of space to maneuver. There must be space to accommodate equipment (e.g., canes, hearing devices) and human and animal aides. Chairs and tables need to be light enough to move around. Tables must be at a comfortable height for a person in a wheelchair.

Not only do the restrooms need to be accessible, but the path to them must be also. Counters, desks, side tables, and other furniture should be free of clutter, flower vases, reading materials, or other items someone with vision or mobility limitations might upset. Reception areas frequently have stacks of clipboards, piles of brochures, and cups filled with pens, any of which could be troublesome. Table 10.1 provides a link to a checklist for evaluating a facility for accessibility.

One additional point about providing a clear path: If you run a session at the *participant's* facility, don't move anything. If you must move an object such as a chair, ask first and then ask if the participant wants you to put it back where it was.

Provide accessible snacks. If you routinely provide snacks and beverages in the test or waiting room, make sure participants will be able to pick them up and open them. A bowl of M&Ms and capped water bottles might be inaccessible food for some participants. Offer to pour coffee or beverages or to place a snack on a plate for participants. Also, prior to their arrival find out if participants have special dietary needs.

10.2.2 **Interacting with blind participants**

To recruit people who are blind, the screening tool must include questions such as the following:

- Will you have an escort?
- Do you need transportation to and from the facility? Do you need assistance getting from public transportation or from a taxi cab?
- Will you bring a service animal? If so, does the animal have special needs? What is its name?
- Do you read Braille?
- Do you use assistive technology (e.g., screen reader, hearing amplifier)?

Make provisions for service animals. Service or guide dogs have never been a problem in our experience, but they do put additional requirements on your interaction. Check with the owner about interacting with the dog. Remember: *When guide dogs are in their harnesses, they are on duty.* They have been trained from an early age that when the harness is on, they are working. When it's off, a service dog can act like a regular dog. Consequently, you need to interact with them differently when they have the harness on. It is best not to talk to them at all. If you do talk to them, use a professional tone of voice and be brief. Never call, pet, or play with a service animal that's in a harness. If you and your staff want to pet the dog after the session, ask its owner.

The service animal will usually lie at its owner's feet during a session. If the session is lengthy, the owner may ask you to take the dog out for a walk. If you don't want to interrupt the session, have a colleague available to walk the dog.

Don't provide food for the dog, but you can ask its owner if it's okay to provide a small container of water. The more water, the more likely the dog will need a walk. Never offer a service animal a treat, food, or water without first asking its owner's permission.

If you run sessions back to back or have multiple participants with service animals in a session, the dogs may interact. Generally, they get along well but keep in mind that they *are* dogs and may object to each other at any point. It's best to keep them separated if possible.

Allow time to print documents in Braille. Organizations that do Braille embossing are not plentiful. Even in large cities, there may be only a few. (See Table 10.1.) In Boston we have been able to get Braille embossing completed in two days, and the cost has not been prohibitive (about $50 per page). Be sure to proof the documents carefully *before* they are sent for printing to avoid a second two-day delay. Ideally, someone should also proof them when they are delivered. If your client is likely to make frequent changes to the wording of documents, explain why you need a hard deadline for final changes.

Consider reading documents to participants rather than printing them in Braille. It can take a blind participant who reads Braille slowly more than 20 minutes to read a two-page embossed document. We have found that blind participants prefer to read documents themselves but often don't object when we ask their permission to read aloud important documents such as consent forms and task scenarios. But keep in mind that you may need to read a document multiple times, slowly and clearly, with expression, and stop frequently to ask if they understand. Alternatively, you may prerecord your reading of the document and either play the audio clip aloud (particularly effective for a group) or through individual equipment such as an MP3 player (which you make available to the participants), enabling individual control and easy repetition.

Check readability of printed documents. Participants with low vision require 16-point or 18-point type to read a document. This is 18-point type. Obviously, you will get fewer characters on a line or a page with type this large. Anecdotally, participants have referred to 18-point type as "a good start"; if your situation enables you to provide larger point sizes, particularly 24-point type, that may be preferable.

INTERVIEW WITH AN EXPERIENCED MODERATOR

Have you ever experienced a conflict between serving the needs of your clients and serving the needs of disabled participants?

Not usually, but sometimes you have to be flexible. Here is an example.

We're halfway through the usability test session at 6:30 in the evening and things are going well. The participant, a woman in her mid-40s with a teaching background and no usable vision, is using a JAWS screen reader to explore a commercial web site. Patty is insightful, articulate, and witty. She is also doing a remarkable job of describing the keyboard shortcuts and macros she's using without interrupting her own navigational process. As a result, I'm furiously taking notes, trying to capture Patty's witticisms, her insights into JAWS use, and her web site–related comments.

Patty's service dog, Harvey, a middle-aged golden retriever, has been snoring contentedly at our feet beneath the computer desk.

The session is taking place in our usability lab at the end of a long day. My colleagues have all gone home. The building is quiet and I'm happy that this session is confirming the findings of the other sessions while providing articulate fodder for the highlight tape and genuine insights into the use of assistive technologies to surf the Web. The participant is smiling and by all indications, we're both having a great time.

And then things change.

Mid-task and nearly mid-sentence, Patty stops, drops a hand to the floor, and gropes for her purse. She rummages through it and produces a well-worn dog leash. Her expression, warm and relaxed just seconds before, is startlingly different: resolute and almost annoyed. "Harvey needs to go walksies," she pronounces.

She moves the leash in my direction, seemingly oblivious to the fact that Harvey is sleeping soundly and that she herself is in the middle of an online purchase.

Briefly taken aback, my mind whirls: Where did this request come from? Why does she seem annoyed? Has the dog somehow been neglected? How does this affect the tasks? Should I stop the recording? Do I leave her here, unattended and alone? Will she be safe? Will she continue to learn things about the web site while I'm gone? Should I insist on her coming with me? Are there legal ramifications? Where is the nearest tree? What if Harvey takes a dislike to me? Takes off? If I continue recording the session, will I have enough video tape to capture the end of the session?

As my thoughts race, I smile, say "Of course!" and take the leash. I also think to myself, When you're lost, stop and ask for directions. *With a calm seriousness that respects but does not exacerbate her obvious concern, I begin to ask questions: "Is Harvey in dire need right now or is this preventive? How does he like to be walked? Will he be ok with my taking him? How long does it usually take? The nearest green space is about fifty yards from the building, and we're about a five-minute walk from the entrance to the building. I'm concerned about leaving you alone here unattended. Your safety is my responsibility. Would you like to come with me at least as far as the parking lot so that you'll be able to offer advice if Harvey gets ideas?" And so forth.*

Together we worked out a solution. As I helped Harvey attend to the local flora, I learned from casual conversation with Patty (who came along) that she had been concerned for days that Harvey might need a walk before the session was through and that I, or my team, might refuse to help him. She was very relieved, as was Harvey, that we were able to resolve the situation easily.

Testing sessions sometimes take off in unexpected directions. Flexibility and a willingness to adapt to participants' needs are crucial. We are hired to serve our clients and to do no harm to participants. Balancing the two sometimes means tipping the scales in one party's favor. Knowing how to do so humanely and deftly takes practice. When in doubt, be direct: Ask the participant and if appropriate ask the client, and make the call that treats the participant with the greatest respect.

For my part, I added a crucial question to our recruitment screeners for disabled candidates: "Are there any concerns you have or accommodations we could make to ensure your comfort and safety while you are participating in this study?" And of course, when accommodating service animals, I now plan to have someone on hand to take them for "walksies." Just in case.

10.2.3 Interacting with deaf and hard-of-hearing participants

Deaf and hard-of-hearing participants present unique challenges. The primary challenge is communication. If you aren't fluent with either American Sign Language or signed English, you will likely require an interpreter. You may also require the use of a CART (Communication Access Real-time Translation) system. Using stenographic machines and computer software, CART aides translate the spoken word into the written word nearly as fast as people can talk; hence, the "real-time." Text is displayed on a laptop computer, monitor, or large screen. CART results in a verbatim (word-for-word) record of all spoken content. (See Table 10.1 for more information.)

Recruiting deaf participants. It's a considerable challenge to find deaf people to participate in usability tests, though it's not as hard as finding people who are both deaf and blind. The best source may be local learning institutions, which often have dedicated disability coordinators who can distribute flyers through email lists of students, alumni, or state- and federal-level disability organizations. During recruiting, you need to ask about the extent of the candidate's ability to hear and his or her use of assistive technologies. You may need to conduct recruitment activities through electronic means such as a teletypewriter (TTY) and email questionnaire. Communicating with people who are deaf or hard of hearing and who are not computer literate may offer additional challenges. During the recruiting process, make sure to ask if they will expect your organization to provide assistive technologies or an interpreter.

Essentially, assistive listening devices (ALDs) are amplifiers. They assist the hard of hearing and sometimes the legally deaf by amplifying sound. But ALDs are not easy to find and are expensive. "Interpreters" are people who are qualified to use sign language (ASL, signed English, or tactile signing) to help the hard of hearing and deaf communicate. They are also used by deaf people who cannot speak clearly enough to be understood. Interpreters are highly skilled and interpreting is exhausting work. Even if a staff member in your organization is a fluent signer, it is unlikely that he or she will be an effective interpreter without special training. You can locate interpreters through the Registry of Interpreters for the Deaf or a state-level organization such as the Commission for the Deaf and Hard of Hearing in your state. In 2007 certified interpreters charged $40 to $80 per hour. Depending on the nature of the communication required, an hour-long session may require multiple interpreters who alternate 15-minute shifts to reduce interpreter fatigue. (See Table 10.1 for resources.)

Moderating with an interpreter present. Interpreters are a channel of communication between people who are deaf and people who can hear. They communicate with the deaf through signed language and with the hearing through speech. In a usability test, the interpreter listens to what the moderator says and signs it to the deaf participant. The interpreter interprets the signs of the participant and tells the moderator what the participant said. Obviously, this arrangement requires some adjustments on your part.

Working with an interpreter can slow the progress of the session, not only because of the translations going on but also because the three parties tend to look at each other while communicating. You do not have to look at the interpreter to hear the interpreter's translation of what the participant signed or to watch the interpreter sign to the participant. In fact, it is considered good practice to ignore the translator as communication is occurring and focus your attention on the participant, who is "speaking" (Henry, 2007). But it takes practice not to watch the interpreter. In a test with participants without limitations, the activity can move very quickly as participants work on tasks and think out loud. In an interpreted session, the thinking aloud is by definition retrospective and filtered through the interpreter. The pace may be slower and it can be difficult to probe about concepts and emotional reactions while they are occurring. This situation is similar to the one in which you have an interpreter because the participant speaks a different language.

In addition, your ability to capture the session through video or audio may require adjustments. If the interpreter is providing the participant's "audio," then he or she might need to wear a microphone while the video capture focuses on the participant.

10.3 **THE ELDERLY**

The elderly are not a homogeneous population! There are large differences in skills and abilities as people move through the decades beyond 60. Furthermore, there are large individual differences among people of the same age. It's beyond the scope of this book to describe in detail the general characteristics of this population. We focus instead on issues that directly influence interacting with usability test participants. Useful sources of information on the elderly are Chisnell, Lee, and Redish (2005) and Tedesco, McNulty, and Tullis (2005).

10.3.1 **Recruiting elders**

Use special strategies. It can be a challenge to find qualified elderly participants who are computer and/or Internet literate. In 2007 the elderly generally don't participate in bulletin boards or chat rooms. They check email less often than younger users and they are wary of strangers both in email and in person. They often screen their calls before answering and they may become agitated if a phone conversation lasts more than 10 or 15 minutes. The best way to find candidates is through personal networks. It makes a huge difference when you can start off a conversation or message with the name of someone the candidate knows. We have had great success using online bulletin boards to contact caregivers or younger people who might know an elderly person who would like to participate in a study.

Visiting or calling senior centers can also be useful. You will find seniors there, but there may be few who have computer literacy, if that's what you require. Try locating the person at the center who teaches a class on computer literacy or the Internet. He or she can then recommend some possible candidates or hand out a flyer for you.

Establish your credibility quickly. This is where a personal reference comes in handy. The elderly are especially wary about sales pitches or disguised sales pitches and about scams. Having a business card from an organization they might recognize helps. Also having a flyer that clearly explains your objective and what the participants will get out of the activity helps reassure them. The candidates can take the flyer with them and read it when they don't feel pressured. If you contact them by phone or email, it is important to let them know how you found them.

Explain the testing situation during recruiting. In addition to the general guidelines about clarity that apply to all the groups in this book, the elderly have some special needs: They, more than other types of participants, often assume that the session will involve a group of participants. They are much more familiar with focus group participation than one-on-one sessions. Make it clear that an individual test is not a group session. It may help to refer to the method as an "interview" rather than a "session" because they are familiar with one-on-one interviews. Finally, remind them to bring their reading glasses if they normally use them.

10.3.2 **Interacting with elders**

When scheduling sessions, keep in mind that the elderly are generally less fatigued in the morning and that they often don't like to drive during rush hour or in the dark. Also, people who have had experience with testing elderly participants say that there is a 20 to 25 percent no-show rate (more than twice that of younger populations). The elderly may forget about the session or have medical problems that take precedence.

Expect them to arrive early and bring a friend. It is not unusual for elderly participants to be 15 to 30 minutes early and to come with a spouse or a friend. Make sure there is place for them both to wait while you get ready or finish an earlier session. The companion will also need a comfortable place to sit and something to read or listen to during the test.

Be polite. Of course, moderators are polite to all participants, but "Please" and "Thank you" are especially important to older people. A common perception of the elderly is that their generation is polite and that younger generations are not. Consequently, they react positively when you thank them and ask questions politely. Also, they are less comfortable with a casual business attitude than younger participants. Do not assume that you can call them by their first names. Address them as Mr. X, Mrs./Miss/Ms. X, or ask them how they like to be addressed.

Minimize interruptions. Some elderly people have difficulty maintaining the focus of their attention when they are interrupted. They are more likely to lose their train of thought or their place in a task or document. This means that it is especially important to wait for a break in the conversation to talk. Be vigilant about watching for signs that the participants need some help to resume what they were doing.

Expect them to tell stories and get off track. There is not much that you can do to prevent these diversions. When they occur, gently move the conversation back to where it needs to be. For example, "Thanks for sharing that. What were we doing?"

Expect them to blame themselves. Self-blame occurs frequently with all participants but especially with the elderly. A lack of technological literacy is a common theme when elders talk among themselves. People older than 65 years in 2007 grew up when computers were not used in schools. They are quick to blame themselves for design failures. You need to provide additional assurance that the tasks are not a test of their abilities. It can be frustrating when they don't seem to hear the reassurance, but it's important to keep providing it.

Consider an older moderator. We are not aware of any research data on this issue, but in our experience, elders are more comfortable if the moderator is closer in age to them. So if you have a choice, an older moderator is preferable.

10.4 PEOPLE WHO HAVE LOW LITERACY SKILLS

One important characteristic of this group that separates it from other groups is its size. It is estimated that 45 million Americans function at the lowest level of literacy. Half of all Americans read no higher than the eighth-grade level (West, 2003). Gribbons (2007) describes how product developers and the human–computer interaction community have ignored this population.

10.4.1 Functional illiteracy

Functional illiterates are able to read and write minimally in their native language but can't perform fundamental tasks needed to live easily in their culture. In the United States, this means they can't fill out an employment application, follow written instructions, read the newspaper, look a word up in a dictionary, or understand a bus schedule.

The causes of functional illiteracy are varied, including cognitive limitations and learning difficulties. The largest group with identified learning difficulties comprises people with dyslexia. But there are also people who were poorly educated and dropped out of school but who have no other measurable limitations. However, many of these people have average or better IQs.

Due to the stigma in our culture against people with low literacy, this condition remains hidden. A test participant is never going to tell you he or she can't read your consent form. They have coping strategies that allow them to survive, such as saying they forgot their glasses and asking you to read it for them. Compounding the problem is that most people overestimate their ability to read. While half of all Americans read no higher than the eighth-grade level, many will say their reading is "good" or "very good" (West, 2000).

10.4.2 **Interacting with functionally illiterate participants**

From the description, you can see that illiteracy is difficult to detect. Most of us who moderate tests have not interacted with many low-literacy participants because they have been almost completely ignored by designers of products that could help them, such as health information web sites. But if full accessibility ever becomes a reality, you will be dealing with people who can't read consent forms or task scenarios.

Our guideline for dealing with this group is to be aware that many people are functionally illiterate and, if you suspect this of a participant, to take appropriate steps:

- Read the consent form slowly to them and probe for their understanding.
- Read the task scenarios to them.
- Accept the fact that they may not be able to think out loud as they work.
- Help them with ratings or questionnaire items.

10.4.3 Testing with low-literacy participants

One of us (BL) has conducted usability tests with people who were selected specifically because they had lower literacy skills. This was a government-sponsored project aimed at helping parents of children with a specific medical condition to communicate with health care providers. The idea was to create a computer system that would allow parents to enter information about medications and their effects. The researchers wanted to ensure that the program would be accessible for both high-literacy and lower-literacy parents, so we ran a usability test with six parents in each category.

The parents had been screened ahead of time using a test for their level of literacy (particularly regarding medical concepts and terms). All of the parents were interviewed before the test and a researcher explained the informed consent in detail. Parents were allowed to provide verbal consent rather than written consent. When participants arrived, the moderator knew the person's approximate literacy level and made accommodations for each person after assessing his or her comfort level with computers and reading comprehension.

Many of the lower-literacy parents freely admitted that they were not good with computers, and, in fact, some of them had used a computer only once or twice. In those cases, we worked slowly, explained interface concepts such as scrolling and drop-down list boxes, and read onscreen instructions out loud. It was tricky to be a Neutral Observer while avoiding pressuring participants to complete tasks when it was unclear whether they could read and understand the interface.

For all participants, we read the task scenarios out loud rather than presenting them on task cards. Likewise, we conducted the post-test interview verbally rather than asking participants to complete a written questionnaire.

It took more effort on our part to make these sessions work. But it was rewarding to be able to make design recommendations that, we believe, will allow low-literacy users to contribute to the health of their children.

10.5 **CHILDREN AND TEENS**

Moderating test sessions with children and teens can be fun and rewarding. If you're working on a product that is aimed specifically at them, then you will have to think about how you recruit them and how you interact with them during the test.

10.5.1 **Grouping children by age**

By "children," we mean people younger than about 15 years. Hanna, Risden, and Alexander (1997) categorize children in one of three age groups.

The first group comprises preschool children ages 2 to 5 years. Usually the most you can do with participants this young is to have them show you what they can do. Their attention span is limited so it's difficult for them to stay on tasks and complete them. It is also hard for them to verbalize their opinions. They won't be able to interact with a product and think out loud at the same time. They are likely to be uncomfortable being alone with a moderator for any length of time.

The second group contains elementary school children ages 6 to 10 years. They can sit and do tasks as they do in school. The older they are, the less self-conscious they are, and the more willing they are to talk. At 6 years, they probably will be uncomfortable being alone with a moderator; at 10 they may not be. They may have difficulty with concurrent thinking aloud.

The third group includes middle school children ages 11 to 14 years. They are more like adults. They are comfortable doing tasks and being alone with a moderator. They are often more technologically savvy than most adults. But they are unpredictable and can't be left alone.

A fourth group of children is those who are 15 to 18 years old.

10.5.2 **Recruiting children and teens**

You can't conduct tests with children younger than 18 without the permission and supervision of their parents. Consequently, several guidelines involve interacting with parents. Some of these guidelines were first described by Hanna and colleagues (1997) and Patel and Paulsen (2002).

Contact and screen children through their parents. Even if you find parents of likely young candidates through a recruiter, you must talk to the parents before you proceed. Parents are very wary of exposing their children to experiences that the parents or children might find uncomfortable. You will need to make the parents feel comfortable and safe and make the case that there is a substantial benefit to their children for being in the study. It helps to indicate that the most important aspect of the child's participation is the benefit to other children and that their children will have an enjoyable and, if possible, educational experience. Incentives might help interest the children, but don't depend on the value of the incentive alone to appeal to parents.

A few more points about recruiting children:

- Don't ask about school performance (grades) unless it is absolutely essential.
- Be very clear and specific about what will happen when the participants come for testing. Parents need to know that they will have to sign a consent form and perhaps a nondisclosure form when bringing them in. Some parents will not consent to having their children videotaped.
- Be prepared to fax or email parents a description of the study and their part in it. They may want to see it before they agree to participate. Use official stationery and include a signature from a key person in your organization.
- If you happen to talk with the child on the phone during recruiting, keep the conversation short. Parents are wary of long conversations. If you happen to talk to the child before talking to the parents, don't ask for their permission to participate. Arrange to speak with a parent.
- Ask specific questions about the children's computing environment and the input devices they use. Sometimes their interaction is limited to game controllers. You may also have to screen for child super users. Some very young children are writing their own programs, doing complex video editing, and building web sites these days.
- Expect to conduct sessions in late afternoon or on weekends when both children and parents are available.
- Find out if a sibling will be present at the test and establish what that sibling will do during the session.
- Provide age-appropriate entertainment in waiting areas. Although showing a movie is relatively easy, consider providing drawing paper, crayons, markers, and other creative and interactive items.
- Encourage teens to come with parents (or parents to come with teens). Teens may ask to come by themselves or to be driven by a friend but if they are under 18, their parents still have to sign the

consent form. Also, teens who come "alone" often come with a crowd of friends who can be unruly and can't be left alone.

- Expect a high no-show rate. Teens frequently arrive late and often don't show up at all.

10.5.3 **Interacting during the session**

Attend to parents and siblings. Children younger than about 9 years will need their parents with them in the session. Keep siblings separated. Young children often are uncomfortable being alone with strangers in a room. You will need to coach the parents about what they can and can't do as visitors. Also, it's rare that a brother or sister close to the same age as the participant can keep from interacting with him or her if they are in the same room. Do everything you can to separate them but always defer to parent's wishes.

Don't ask children if they want to do a task. They are much more likely than adults to say "no." A useful prompt is, "Now, I need you to . . ." (Hanna et al., 1997)

Prompt to keep children on task. Their attention span may be limited. If they start looking around, say "I need you to keep trying this for 5 more minutes, then we can try something else" or "I want to see just how much you can do—let's try some more." Once a child says, "I don't want to do this anymore," it is very difficult to continue (Hanna et al., 1997).

Look for nonverbal signs of feelings. Watch children's body language and what they look at. Squirming and looking around the room often indicate fatigue. If you suspect that they are becoming uncomfortable, take a break and talk about a neutral topic until you get a sense for their emotional state.

10.6 **PEOPLE FROM OTHER CULTURES**

Almost all of the research and literature on usability testing is from tests in which the participants and the moderator are from the same culture, typically from the United States and Western Europe. Until recently, the impact of culture on product design and usability studies has been largely ignored (Marcus, 2006).

Over the past decade, access to the Internet has spread to many countries, especially those in Asia. Usability labs have started to appear in India and Korea. Consequently, it is becoming more common for moderators and test participants to be from different cultures.

10.6.1 **Interacting with participants from the same and other cultures**

Several useful articles in the literature address the differences in assumptions about interacting in usability studies when it happens between people from different cultures. An excellent case history about conducting usability studies in China by Elaine Ann can be found in chapter 5 of Courage and Baxter's book *Understanding Your Users* (2005). Ann points out that people in China would have difficulty with the mixture of the two roles of a moderator, the Leader and the Neutral Observer. In China, business relationships are built on friendships. Consequently, Western moderators maintaining what they might view as a professional, businesslike demeanor might be viewed in China as aloof and not to be trusted.

There has been some recent empirical research on the interaction of moderators with participants from different cultures and its effect on the interactions in a usability test. Evers (2004) conducted think-aloud tests and post-test interviews with a sample of 130 high school students from England, North America, the Netherlands, and Japan. The moderator was English. The Japanese students had the most difficulty with the think-aloud sessions. They felt uncomfortable speaking out loud about their thoughts and seemed to feel insecure because they could not confer with others to reach a common opinion. The English also needed reassurance before feeling comfortable with thinking out loud. The interview responses of the North Americans were often inconsistent with behavior observed during the tasks. Evers felt participants were trying to give the "right answer" rather than report their true feelings in relation to the web site being tested. The Japanese participants seemed very comfortable in voicing positive as well as negative opinions of the web site.

Even when moderators and participants are from the same culture, reactions can be different from what you would expect in the United States. Yeo (2001) found that Malaysian participants were less likely to be critical of a user interface when the moderator is a Malaysian person they know versus a stranger.

The study by Vatrapu and Parez-Quinones (2006) that is described in the accompanying "What the Research Says" provides additional and strong evidence that fewer usability problems may be uncovered when the moderator and participant come from different cultures, even if they attend the same university.

WHAT THE RESEARCH SAYS

Do cultural differences affect moderator interactions?

This study investigated whether test participants change the way they interact with a moderator when the participants are from the same or a different culture (Vatrapu & Parez-Quinones, 2006). The authors recruited a sample of students at Virginia Tech who had come to school from India. They gave the students a test that measured their acculturation level and they chose participants who were low in acculturation, meaning that they had not yet absorbed much about American culture.

The sample of students was then separated into two groups in preparation for a usability test. One group had a moderator who was Indian and one group had a moderator who was Anglo-American. The moderators were not told about the other group and were both given the same moderator's test script to review. The scripts were identical except that the Indian moderator's script had the following insert:

"I am from Andhra Pradesh. Did you attend the Indian Film Festival held recently? I watched three movies."

In the second phase of the study, four graduate students in usability science reviewed the videotapes from the sessions and scored the post-test interviews by identifying usability problems, suggestions, and positive, critical, and cultural comments. Two raters each scored either the sessions from the Indian moderator or those from the Anglo-American moderator.

The results showed that when the Indian participants had an Indian moderator, they made more critical comments about the product, fewer positive ones, and more suggestions. In addition, they rated the product as more difficult to use and their discussion with the moderator uncovered more usability problems. All of these measures were statistically significant.

This study shows that, at least for the Indian culture, participants are more forthcoming when they have a moderator from the Indian culture and, consequently, more usability problems are uncovered. On the other hand, when the moderator is Anglo-American, Indian participants are more reluctant to be critical of the product so fewer problems are uncovered.

Vatrapu, R., & Parez-Quinones, M. (2006). Culture and usability evaluation: The effects of culture in structured interviews. *Journal of Usability Studies*, 156–170.

10.6.2 **Adapting your techniques**

Apala Chavan of Human Factors International has reported success with the "Bollywood technique" (Schaffer, 2002). Bollywood is the Hollywood of India where they make movies that have intense and elaborate plots. In the Bollywood technique, the moderator describes a task scenario in the context of a dramatic situation. Chavan talks about a test in which one of the tasks was to make a train reservation. A typical scenario that just asked participants to make a reservation was not succeeding in that the participants did not seem motivated and were not very explicit about the reason. She then described a scenario in which the participant's beautiful, young, and innocent niece is about to be married. But suddenly he gets news that the prospective groom is a member of the underground. He is a criminal and he is married! The participant has the evidence and must immediately book a train ticket for himself and the groom's current wife to Bangalore.

The participants seemed to get involved in this dramatic scenario immediately and made many insightful and critical comments. Chavan believes that these kinds of dramatic scenario work well, at least in that part of India.

Of course, we cannot create a different set of guidelines for every cultural combination of moderators and participants. We urge moderators to be aware of cultural influences, to learn about other cultures, and to consider them when recruiting and interpreting usability tests. There is some research too that shows that culture matters, and there are undoubtedly many more cultural differences yet to be discovered that affect usability testing.

11

Integrating the videos

One of the reasons that there is so little in the literature about interacting with participants in usability tests is the difficulty of describing in words the subtleties of communication. We have talked about body language, nonverbal cues, facial expressions, tone of voice, and so on. Seeing these portrayed in real testing situations reinforces the golden rules of moderating and illustrates the various roles of a moderator.

When we conceived the UPA tutorial that eventually led to this book, we suspected that its success would depend on having interesting videos to show both good and poor practices. The attendees confirmed our expectation; the videos became the focus of many discussions and were praised in evaluations of the tutorials.

Consequently, we have made the tutorial videos available on this book's web site, *www.mkp.com/moderatingtests*. We've refilmed several of them, so even if you attended one of our tutorials, you will see some new material.

11.1 **ABOUT THE VIDEOS**

Before you look at the seven videos, we want to explain their purpose and provide some advice about how to view them.

11.1.1 **The test session videos**

Six videos show test sessions in progress. They are not highlight videos; rather, each one focuses on one 5- to 7-minute segment of a typical test session. Two videos focus on the pretest instructions, two focus on interacting while tasks are being performed, one focuses on the post-test interview, and one illustrates a remote test.

Each video shows several situations that illustrate the major points, but not all of the situations, in this book. The people in the videos are acting; most of them are usability specialists. The events they portray are based on real events we have experienced. Some of the videos include poor practices, which we created for the tutorial to generate discussion.

Because the videos portray typical events, you may find some of them "slow." We have chosen realistic fidelity over highlight summaries. There is no way, for example, to illustrate the gradual buildup of frustration in a participant without taking the time to show it. There is no way to realistically illustrate a moderator letting a participant talk and work for a few minutes without showing it. The fact that you might treat the situation differently is exactly the point. Consider your reaction carefully and think about why another approach might be more effective than what the video shows. Discussing your reaction with your colleagues is a good way to incorporate the principles of interacting into your testing practice.

Each video has an accompanying panel discussion by a group of experts. We invited them to our facility to watch the videos while we recorded their reactions. We asked for their views of the videos because we knew that they would bring up issues and options that we hadn't thought of. Their reactions to one another's comments are a valuable supplement to the material in this book.

11.1.2 **Use of the videos**

We suggest that you watch a video first, then look at the section of the panel discussion about that video. There is no need to watch them in order from first to sixth. For example, if you are interested only in remote testing, we suggest that you read chapters 3 and 4 to understand the golden rules and chapter 8 about remote testing. Then watch Video 5 on remote testing and the panel's reaction to it. Finally, you might look at the section in this chapter about Video 5, which describes the key events in the video and our comments on them.

We have learned from our tutorials that viewers enjoy the videos more when they watch them in a group. The different reactions of the group members enhance the value of the videos. Talking about the events as you see them also helps make the rules of interacting more real.

11.2 CONTENT OF THE VIDEOS

The following description of each video includes an introduction about its purpose and a table that describes the actions in sequence and offers comments about their significance.

- Video 1: Pretest briefing with a checklist.
- Video 2: Pretest briefing following a script.
- Video 3: Interacting during the session, example 1—illustrates how to deal with direct questions from the participant and how to deal with the administrator's anxiety over the participant's stress.
- Video 4: Interacting during the session, example 2—focuses on when and how to give participants assistance on a task without introducing bias.
- Video 5: Interacting in a remote testing situation—illustrates how to set up and conduct a remote session, as well as the difficulties of not being able to see the participant.
- Video 6: Post-test interview—illustrates how to get the most out of the valuable time at the end of a session, and how to deal with questions from the participant.

11.2.1 Video 1: Pretest briefing with a checklist

The purpose of the video is to illustrate the way an experienced moderator typically gives the instructions at the beginning of a session. The moderator and the participant in this video are sitting facing each other. The participant is sitting on a couch on one side of the Bentley College usability lab. We often give pretest instructions there because it seems more relaxing than sitting in front of a computer screen. The moderator does not read the instructions but rather follows a checklist on his clipboard. The moderator's demeanor is relaxed but professional. Figure 11.1 highlights the events and the issues they illustrate in this video.

11.2.2 Video 2: Pretest briefing following a script

This video has the same scope as Video 1 but with two differences: (1) the moderator is inexperienced and (2) he reads most of the instructions from a script. Compare and contrast this option with the experienced moderator's style and coverage of issues in Video 1. Reading a script is a perfectly acceptable practice when done well. In most research studies, researchers use a script to make sure that every test participant hears exactly the same instructions. In our experience,

What happens	Comments
Moderator introduces himself and thanks participant for coming.	Helps to create a connection with participant.
Moderator explains what will happen during the session and how long it will last.	Part of moderator's responsibility for informed consent.
Moderator explains that there will be tasks and that he will not help her with them "right away."	Sets expectations about what kinds of help participant can expect.
Moderator explains that not all tasks need to be completed.	Anticipates stopping a task before participant has completed it. Tries to avoid having participant blame herself for failure.
Moderator tells participant what thinking out loud is.	This explanation is all some moderators say about thinking aloud.
Moderator models thinking aloud: having an expectation ("I am looking for instructions on how to open the stapler") and a feeling ("That was easy") along with describing actions ("I think I will just shut it").	Moderator shows participant that she should talk about expectations and feelings in addition to describing actions.
Moderator asks participant to practice thinking out loud.	Moderator is focused on his checklist, not on participant. He may be anticipating what he will do next.
Moderator explains informed consent form; recording data and what will be done with it; there are people watching behind the one-way mirror. Urges participant not to blame herself.	Should moderator have done the form earlier in the instructions? Moderator does not fidget while participant signs the form.
Moderator tells participant about limitations of pointing at the screen.	A point that is easy to forget during the tasks.
Moderator tells participant she can ask for a break at any time.	One of participant's rights.
Moderator asks participant if she has any questions.	Required by informed consent.
Did moderator forget any items?	He did not inform participant that she could withdraw without penalty at any time.

■ **FIGURE 11.1** Video 1: Pretest briefing with a checklist.

usability test participants don't have a problem when you read instructions as long as they are told why the instructions are being read and as long as the moderator still pays attention to the participant. In this video, the moderator tells the participant that he is reading because he wants to make sure he gives everyone the same instructions. But as you will see, the atmosphere in this video is quite different from that in Video 1. This moderator is much more hesitant but much more thorough than the moderator in Video 1. Figure 11.2 highlights the events in this video and the issues they illustrate.

What happens	Comment
Moderator greets participant and introduces himself.	Looks at participant to make a connection. Voice is pleasant.
Moderator tells participant why he will be reading the script.	Is "apologize" the right word here? What about just stating why he is reading?
Moderator tells participant the purpose of the facility and introduces the concept of testing for usability.	As moderator begins to read, he's making a point of looking up at participant regularly. He does stumble a bit, showing his inexperience.
Moderator tells participant about session's three parts.	Notice that after stumbling, moderator looks up less than before.
Moderator tells participant about the consent form.	Notice that now participant is looking at moderator's script as he looks down at it.
Moderator notes that there may be people watching and participant reacts. Moderator is aware enough to see participant's reaction and ask her about it: "I notice you had a reaction to that."	Moderator seems a bit nervous answering participant's question: "There are about two developers." But note that when participant interrupts moderator, he stops and listens to what she's saying.
Moderator explains informed consent and confidentiality issues before he gives participant the form.	Moderator is much more thorough about these issues than moderator in Video 1.
Moderator makes it clear that participant should take her time and should sign the form if she is comfortable with it.	Moderator does not fidget as participant reads and signs the form.
Moderator asks participant if she has any questions after she signs the form.	This is an important part of informed consent.
Moderator carefully tells participant not to blame herself and why.	He seems genuine about this.
	At this point you get two impressions: moderator is a bit unsure of himself but is genuinely concerned that participant gets thorough instructions.
Moderator models thinking aloud using a stapler.	Moderator stops reading and his explanation is much smoother. The pace of his speech increases.
As moderator models, he expresses feelings, expectations, and judgments of ease of use.	Moderator makes it clear that thinking aloud is more than just reporting actions.
Participant practices thinking aloud.	Moderator watches what participant's doing rather than looking at his script to see what's next.
Moderator explains his roles to participant.	Moderator seems more relaxed now and his speech flows more smoothly.
Moderator tells participant that he may stop tasks before she is done and why.	
Moderator explains how to start and stop tasks.	
Moderator ends the instructions by asking if participant has any questions.	Notice that participant uses moderator's name, perhaps to indicate her connection. Moderator has not used participant's name.

■ **FIGURE 11.2** Video 2: Pretest briefing following a script.

11.2.3 **Video 3: Interacting during the session, example 1**

This is one of two videos that deal with events that occur while participants are working on tasks. This video presents a situation in which the participant is increasingly annoyed while working on a task. The moderator handles some situations well but he is less than perfect in others. Figure 11.3 describes the events in this video and offers comments on them.

11.2.4 **Video 4: Interacting during the session, example 2**

This video also illustrates events that occur while the participant is attempting tasks. The participant has difficulty completing the task and the moderator gives her an assist. This moderator has a less interactive style than the moderator in Video 3. Figure 11.4 describes the events in this video and offers comments on them.

11.2.5 **Video 5: Interacting in a remote testing situation**

This video starts with the final few moments before a remote session begins and then continues with events that illustrate some issues that arise in this type of session. The moderator is using a phone-conferencing system and Internet-based web-conferencing that integrates the phone-conferencing audio into the recording. The test participant and the observers have been sent the phone number, the conference ID, and the web-conferencing URL and ID. Figure 11.5 describes the events in this video and offers comments on them.

11.2.6 **Video 6: Post-test interview**

The purpose of this video is to depict some key events that occur during a post-test interview at the end of a diagnostic test. In making this video, we recognized that there may well be more probing about events that occurred during the session. But we could not make that meaningful without showing the earlier events. Figure 11.6 describes the events in this video and offers comments on them.

11.3 **THE FUTURE OF USABILITY TESTING**

As we finish writing this book, we see a time of change in our profession. There are several trends that we believe will affect the way we

What happens	Comment
Moderator hands participant the task scenario and asks him to read "go ahead and read task 1 for me."	Moderator's tone is friendly.
Participant begins looking for a way to do task. Moderator asks, "Tell me what you are thinking." (Notice that the moderator keeps tapping his pen on the clipboard. A habit he is not aware of.)	Moderator is trying to move participant away from just reporting actions by asking about "thinking." Moderator's action works. Participant explains his strategy. He has moved to more meaningful comments.
Participant says, "I don't even know where to start."	This comment seems to be a hint that participant wants help. He's been working on the task for less than a minute. Moderator ignores comment and asks what the participant would be looking for.
Participant repeats statement that he has no idea where the control is and would call customer support.	Moderator tells participant that he wants him to try to find the solution on the site itself.
Participant blames himself, "That was my fault."	Moderator assumes that participant is not unusually stressed at this point. Moderator continues to take notes.
Participant blames himself again.	Moderator tells him he is doing fine.
Participant remarks that computer response time is, "really slow."	Moderator asks if participant would expect it to be slow.
Participant says that the route takes too much time.	Moderator asks him to find a faster route. He wants participant to explore additional options.
Participant again says he has no idea how to do this task.	Moderator says, "I would like you to work a little harder—a little longer. There is a way to do it." Moderator catches himself in mid-sentence suggesting that participant has not been working hard and quickly changes it to "working longer." Moderator adds that there is a way to do the task in case participant thinks this is a trick task.
Moderator says, "And don't worry. Everyone has difficulty with that particular task."	Did moderator have to tell participant about what other participants have done? Did he make this up to make participant feel better? What are the implications of this statement? Will this lower participant's ratings of usability? How else could moderator encourage this participant?

■ **FIGURE 11.3** Video 3: Interacting during the session—Example 1.

interact with participants in the future. We expect that the next few years will bring more challenges than we have had for the past 15 or so. Usability evaluation is no longer a choice among three or four standard methods. The trend is toward a large toolkit from which we pick a combination of methodologies tailored to each situation.

What happens	Comment
Participant reads the scenario.	Moderator sits slightly behind participant.
Participant looks through menus in a random pattern.	Moderator watches and takes notes.
Participant looks for a "sound" option, but has no idea where it might be.	Moderator waits. Would you say something here? What?
Participant's voice and body language begin to show some negative emotion. She looks at moderator as if she is asking for guidance. Says, "You would expect to find this easy."	Moderator continues to take notes.
Participant sighs.	What is participant conveying with her sigh? What would you do?
Participant says she has looked at all of the obvious places. Moderator says, "You seem frustrated."	What is participant saying here? There is a feeling, but is it frustration? Could it be described with other terms?
Participant says, "Yah, well, frustrated (but as a question), I guess . . ."	Angry? Is moderator putting a label on the emotion instead of letting participant do it?
Participant says, "Can you help me?" Moderator looks down and says, "I would like you to keep trying."	Is additional justification needed to ask participant to continue working? What is gained by spending more time on this task? What would you do?
Participant continues to sigh and is making no progress.	Moderator is clearly watching participant closely, but chooses not to intervene.
Moderator provides an assist, "Why don't you go back to the Edit menu and look at Preferences?"	This level 3 assist gives participant the next step.
Participant completes task and says, "That was frustrating, I guess."	What does "I guess" mean here? Is participant upset at having moderator label her feeling? Is she expressing her feeling? What do you think?

■ **FIGURE 11.4** Video 4: interacting during the session—Example 2.

As professionals, we can't be isolated. National and local professional meetings are essential for keeping up with the latest methods. Here are some trends that we see.

11.3.1 **Current trends**

First, there's the movement away from testing in the laboratory. Remote testing has so many advantages that we expect it to continue to grow in popularity. This trend will require that you learn to listen for clues to participants' feelings and to communicate with the tone and quality of your voice.

What happens	Comment
Moderator goes through his checklist before session starts. Moderator calls a key observer to make sure she is going to attend the session. He reminds her to mute her phone.	The call is part of a strategy to get developers to attend sessions. It's important that observers mute their phones.
Moderator starts the web conference. He calls the phone-conferencing system and puts in his ID to start the conference. No one is there yet.	Taking these actions before the session starts saves valuable time that can be used with participant.
Moderator hears participant join the phone conference and greets him.	By asking whether this is still a good time for the session, moderator is empowering participant to state his preferences.
Moderator checks to make sure participant has the list of task scenarios he emailed and checks that participant has joined the web conference. He also tells participant that a colleague may join the conference.	Telling participant about observers is an important component of informed consent.
Moderator shows participant a slide with squares that indicate how much of moderator's screen he can see.	This is a quick way to make sure participant is using the same resolution and screen size as moderator.
Participant wants to know whether moderator can get into his computer.	Participants don't always know the limits of desktop sharing.
Moderator explains the purpose of the test and reviews the informed consent principles.	It's just as important to explain informed consent in a remote test as in a local one, even when they say they have read it.
Moderator maximizes the application and then checks to see that participant is seeing it.	The pause between when moderator and participants see a screen change is normal.
Moderator gives thinking out loud instructions.	Moderator does not model thinking out loud or train participant. Do you think the brief instruction is sufficient?
Participant starts working on a task.	Moderator is listening carefully to what is being said and how it is said.
The break in the video indicates that the action has moved to the end of the session.	
Moderator thanks participant for his time and goes over the method of payment.	It is important to thank participant and sound like you mean it.
Participant asks if he will get the results.	Moderator explains why that will not be possible. Does this explanation sound evasive to you or is it OK?

■ **FIGURE 11.5** Video 5: Interacting in a remote testing situation.

Second, there is more interest than ever in testing in users' environments. Alternative moderator–participant arrangements will become more common and you will have to be creative in dealing with distractions from ringing phones and interruptions from participants' colleagues and family.

What happens	Comment
The video begins with participant saying she is done.	Moderator has been in the observation room while participant worked on tasks.
Moderator enters the test room and tells participant she can relax. Moderator gives participant a rating form to fill out.	Notice the touch and participant's subtle startled reaction. Touching is a definite "Don't do."
Participant asks a question about the product. She wants to know how the different reporting options work. Moderator seems to be a bit on the spot and says, "We can go over that at the end."	What else could moderator say here? What do you think about moderator's use of "we" here instead of "I can show you at the end?"
Moderator asks whether participant has any other questions or if she needs a drink.	Because moderator put off participant when she did ask a question, should she have asked about other questions here? What, if anything, is gained by asking for additional questions at this point?
Moderator leaves the room while participant completes the ratings.	Moderator is giving participant space to take her time and not be observed. Is there an implicit agreement here that what she writes will be kept private?
Participant finishes and moderator returns. Moderator begins to discuss participant's ratings. "Can you tell me why you gave it a 5?"	Going over the ratings allows observers or people viewing the recording to hear what the ratings were and why. But does this violate the unstated assumption that the ratings are private? Should moderator have told participant that she would go over the ratings with her?
In the discussion of the second rating, participant talks about the reporting function but not its usability. Moderator specifically asks her to comment on its ease or difficulty of use.	Participants often talk about functionality rather than usability.
Participant says she had a hard time figuring out which report menu to use. Moderator relates this comment to the next rating of "Clarity."	Moderator interjects her assumption that the reporting problem relates to "Clarity" rather than the "Usefulness" rating they had been discussing. How about saying: "Share some more with me about the reporting function."
Participant talks about terminology issues.	Moderator misses an opportunity to ask about terms that might have been confusing.
Participant asks if the developers are "geeky programmers." Moderator's facial expression seems to convey uneasiness with this direct question. She says, again, that "we" will get back to that in a few minutes.	Telling participant that they will discuss her question later begins to seem like a put-off when used more than once. Other possible responses are "Tell me what you think about the usability of the product" or "Help me to understand why you think they might be 'geeky'?"
Moderator switches the conversation to participant's original concern about the report menus. Moderator asks participant to look at the two menus and "see if you can see any differences, if there are any."	Moderator lets participant discover the differences rather than just telling her what they are.
Moderator asks how participant would design the reporting function. Participant hesitates a while and says she would make it a one-click operation.	It's rare that participants provide useful design ideas. Design is not their job and, sometimes, they feel put on the spot by being asked.
Moderator asks for participant's top three likes and dislikes.	Moderator specifically asks about "ease or difficulty of use" to avoid having her talk about functionality. Also "ease or difficulty of use" is better than jargon—"usability."

■ **FIGURE 11.6** Video 6: Post-test interview.

Third, the movement toward agile development methods will force usability practitioners toward faster tests of smaller pieces of functionality. *Agile development* is a method for speeding up the development process and creating better products. It consists of a dedicated development team that works together by designing small pieces of the product during short intervals—for example, three weeks. During those weeks, the code is tested iteratively until it is all working correctly. Then the next piece is designed. Usability test sessions as part of this method tend to be shorter, and there is pressure to reduce the time between conducting the session and communicating the results. If the integration of usability into agile processes continues, it will mean that the whole development team will work together more closely, which will force you to put more emphasis on managing observers of test sessions.

Fourth, the Rapid Iterative Test and Evaluation (RITE) method focuses on quickly finding the cause of issues and probing to understand if design changes have solved the issues they were intended to fix. This method may require you to become more assertive at probing for mental models and user-interface concepts.

11.3.2 **What's next?**

We believe that our ten golden rules for interacting will remain relevant in their essentials for some time because they are based largely on principles of human communication. But it's inevitable that the details will evolve and that the relative importance of the rules will change. We look forward to this evolution.

We hope more practitioners will focus on interaction because it is so important to the validity and reliability of usability evaluation, and because so much of our knowledge is based on unsystematic observation rather than solid research.

Finally, we're confident that changes in the ways we interact with test participants will move toward bringing users into closer partnership with product teams. The people orientation of our profession—its greatest strength—will continue to motivate all moderators of usability tests. That is why, after all, we were drawn to it.

References

Books

Anderson, J. (2000). *Cognitive Psychology and Its Implications*, 5th ed. New York: Worth Publishers.

Barnum, C. (2002). *Usability Testing and Research*. New York: Longman Publishers.

Courage, C., & Baxter, K. (2006). *Understanding Your Users: A Practical Guide to User Requirements*. San Francisco: Morgan Kaufmann.

Dumas, J. (1988). *Designing User Interfaces for Software*. Englewood Cliffs, NJ: Prentice-Hall.

Dumas, J., & Redish, J. (1999). *A Practical Guide to Usability Testing* (rev. ed.). London: Intellect.

Dumas, J., & Redish, J. (1993). *A Practical Guide To Usability Testing*, 1st ed. Norwood, NJ: Ablex Publications, Inc.

Mitchell, C. (2007). *Effective Techniques for Dealing with Highly Resistant Clients*, 2nd ed. Johnson City, TN: Clifton Mitchell Publishers.

Nielsen, J. (1993). *Usability Engineering*. New York: Academic Press.

Rubin, J. (1994). *Handbook of Usability Testing: How to Plan, Design, and Conduct Effective Tests*. New York: John Wiley & Sons.

Sales, B. D., & Folkman, S. (2000). *Ethics in Research with Human Participants*. Washington, DC: American Psychological Association.

Snyder, C. (2003). *Paper Prototyping: The Fast and Easy Way to Design and Refine User Interfaces*. Boston: Elsevier/Morgan Kaufmann.

Note: We have divided the references into three sections: books, articles, and Internet resources.

West, D. (2003). *State and Federal E-Government in the United States.* Providence, RI: Center for Public Policy, Brown University.

Articles

Barker, R. T., & Biers, D. W. (1994). Software usability testing: Do user self-consciousness and the laboratory environment make any difference? *Proceedings of the Human Factors Society, 38th Annual Meeting,* 1131–1134.

Beauregard, R. (2005). One of five presentations given at a session "10 Minute Talks on Usability Issues." Usability Professionals' Association annual meeting, Montreal, Quebec.

Birru, M. S., Monaco, V. M., Charles L., Drew, H., Njie, V., Bierria, T., Detlefsen, E., & Steinman, R. A. (2004). Internet usage by low-literacy adults seeking health information: An observational analysis. *Journal of Medical Internet Research,* online publication at: *http://www.pubmedcentral.nih.gov/articlerender.fcgi? artid=1550604.*

Boren, M., & Ramey, J. (2000). Thinking aloud: Reconciling theory and practice. *IEEE Transactions on Professional Communication,* 43(3), 261–278.

Branch, J. (2000). The trouble with think alouds: Generating data using concurrent verbal protocols. *Proceedings of the 28th Annual Conference of the Canadian Association for Information Science.* Also at *www.slis.ualberta.ca/cais2000/branch.htm.*

Brooke, J. (1996). SUS: A quick and dirty usability scale. In: P. W. Jordan, B. Thomas, B. A. Weerdmeester, & I. L. McClelland (eds.). *Usability Evaluation in Industry,* 189–194. London: Taylor & Francis.

Brush, A., Ames, M., & Davis, J. (2004). A comparison of synchronous remote and local usability studies for an expert interface. *Proceedings of CHI 2004,* 1179–1182.

Evers, V. (2004). Cross-cultural applicability of user evaluation methods: A case study amongst Japanese, North-American, English and Dutch Users. *Proceedings of CHI 2002,* 740–741.

Gribbons, W. (2007). Universal accessibility and functionally illiterate populations: Implications for HCI, design, and testing. In J. Jacko & A. Sears (eds.), *The Human–Computer Interaction Handbook,* 2nd ed. Mahwah, NJ: Lawrence Erlbaum Associates.

Hanna, L., Risden, K., & Alexander, K. (1997). Guidelines for usability testing with children. *Interactions*, September–October, 9–14. New York: ACM Press.

Hass, C. (2004). *Conducting Usability Research with Participants with Disabilities: A Practitioner's Handbook*. Washington, DC: American Institutes for Research.

Hawley, M., & Dumas, J. (2006). Making sense of remote usability testing: Setup and vendor options. *Proceedings of the Usability Professionals' Association Annual Meeting*, 1–12.

Lesaigle, E. M., & Biers, D. W. (2000). Effect of type of information on real-time usability evaluation: Implications for remote usability testing. *Proceedings of the IEA 2000/HFES 2000 Congress*, 6, 585–588.

Loring, B., & Patel, M. (2001). Handling awkward usability testing situations. *Proceedings of the Human Factors and Ergonomics Society, 45th Annual Meeting*, Santa Monica, CA, 1–5.

Marcus, A. (2006). Culture: Wanted alive or dead? *Journal of Usability Studies*, 1(2), 62–63.

Medlock, M. C., Wixon, D., Terrano, M., Romero, R., & Fulton, B. (2002). Using the RITE method to improve products: A definition and a case study. *Proceedings of the Usability Professionals' Association Annual Meeting*, 1–5.

Miller, L. (2006). Interaction designers and agile development: A partnership. *Proceedings of the Usability Professionals' Association Annual Meeting*, 1–4.

Patel, M., & Paulsen, C. A. (2002). Strategies for recruiting children for usability tests. *Proceedings of the Usability Professionals' Association Annual Meeting*, 1–5.

Roberts, T., & Moran, T. (1982). Evaluation of text editors. *Proceedings of Human Factors in Computer Systems*, 136–141. New York: ACM Press.

Rosenbaum, S., Rohn, J., & Humburg, J. (2000). A toolkit for strategic usability: Results from workshops, panels, and surveys. *Proceedings of CHI 2000*, 337–344.

Tedesco, D., McNulty, M., & Tullis, T. (2005). Usability testing with older adults. *Proceedings of the Usability Professionals' Association Annual Meeting*, 1–5.

Tullis, T., & Stetson, J. (2004). A comparison of questionnaires for assessing website usability. *Proceedings of the Usability Professionals' Association Annual Meeting*, 1–12.

Vatrapu, R., & Parez-Quinones, M. (2006). Culture and usability evaluation: The effects of culture in structured interviews. *Journal of Usability Studies*, 156–170.

Waters, S., Carswell, M., Stephens, E., & Selwitz, A. (2001). Research ethics meets usability testing. *Ergonomics in Design*, 14–20.

Yeo, W. A. (2001). Global-software development life cycle: An exploratory study. *Proceedings of CHI 2001*, 104–111.

Internet resources

Ambler, S. (2004). Agile Usability: User Experience Activities on Agile Development Projects, online at *www.agilemodeling.com/essays/agileUsability.htm*

The American Psychological Association Code of Ethics, online at *www.pa.org/ethics/code2002.html*

Chisnell, D., Lee, A., & Redish, J. (2005). Recruiting and Working with Older Participants. AARP, online at *www.aarp.org/olderwiserwired/oww-features/Articles/a2004-03-03-recruiting-participants.html*

Hartman, M. (2005). People with and without disabilities: Interacting and communicating, online at *eeo.gsfc.nasa.gov/disability/publications.html#vision*

Henry, S. (2007). Just Ask: Integrating Accessibility Throughout Design, online at *www.uiaccess.com/accessucd/index.html*

The Human Factors and Ergonomics Society Code of Ethics, online at *www.hfes.org/web/AboutHFES/ethics.html*

Nielsen Norman Group (2003). 233 Tips and Tricks for Recruiting Users as Participants in Usability Studies, online at *www.nngroup.com/reports/tips/recruiting/*

Ramey, J. (2001). Methods for successful "Thinking Out Loud" procedures, online at *http://www.stcsig.org/usability/topics/articles/tt-think_outloud_proc.html*

Schaffer, E. (2002). The Bollywood Technique. *www.humanfactors. com/downloads/jun02.asp*

The Usability Professionals' Association Code of Ethics, online at *www.upassoc.org/about_upa/structure_and_org_chart/code_of_ conduct.html*

Index